URBAN ARCHITECTURAL DECORATION

城市建筑装饰

香港理工国际出版社 主编

中国·武汉

EMBELLISH CITY WITH ARTS

PREFACE

CONCRETIONARY POETRY

Architecture embellishment is a kind of poetry.

There are numerous high-rises in the city which are just products of rapid development.But architecture embellishments are different from them.They are the master of public realm. They are geniuses with light and shadow.They get rid of gravity with unique rhythm when they are dancing above and extending free.Some of them are Just up from a simple base.Then without any foreshadowings or transitions, they are towering high in the air, overlooking the city.Some of them are unique with small volume and colorful façade, representing the identity of the city and the feature of the era.

There is no convention between fiction and fact. Maybe traditional exterior wall is not heavy. Maybe neat structure is not rigid. Maybe huge volume is not clumsy. Under the wild, vertical and strong line, there is something tactful, intricate or exquisite.Under the smooth, luxury and fluid façade, there is something cool, concise or extended.Even the light is seemed especially delicate and graceful, because of the elegant architecture embellishments.

However, the beauty of architecture embellishments is not only on the surface, but also inside them. Architecture embellishment is a kind of poetry.It is a piece of poetry which enshrined love and grace.

Every project in this book, is like an exquisite silver button which decorating the city cloak with shining light.Only if you feel the elegant of the architecture embellishment and enjoy the poetry, you will see the real beauty of the urban and touch the deep part.There must be a proper architecture embellishment which suit the public realm and even the whole city. Then it will shine and attract every citizens.

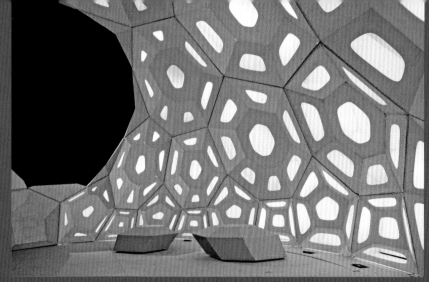

前言

凝固的诗情

那是怎样的一份诗情,凝固成这琼楼玉宇?

水泥丛林内高楼大厦无数,唯独这些空间的主角、光影的精灵,以特有的节奏,摆脱重力,凌空飞舞,自由伸展。有的从一个平凡的基点拔地而起,不需要铺垫和过渡,轻易地甩下世俗和浮尘,兀立云端,俯视全城;有的造型新颖、体量小巧、色彩张扬,寄托着某种如歌情怀,彰显着某个时代特点。

虚实之间并无常理,传统的外墙并不一定厚重,规整的结构并不一定死板,庞大的体量并不一定笨拙。在粗犷、垂直、强劲的线条下,也可以是委婉、曲折与细碎;在光滑、华美、流利的外表下,也可以是冷峻、精炼与延展。就连光线,也因美妙的建筑,显得格外精致、缠绵,甚至绰约。

然而,建筑之美不可浮于表面,它需要内涵与外延。优秀的建筑就像一首诗,不但格式整齐,字字珠玑,自由流畅,却又显得婉约蕴藉,余味隽永。

本书中的建筑如同一枚枚精致的银扣,缀在都市大氅上,不分昼夜地熠熠闪光。你只有亲眼看到细致的装饰,感受凝固的诗情,才会惊讶城市的繁华和丰富,叹服城市的高雅与深邃。朦胧之中,总会有一抹灿烂与现实重叠,袅袅卷舒,交相辉映,有如晕开的墨花,也似浮在水上的轻波涟漪,令人陶醉不已。

CONTENTS

008 LEAF CHAPEL 树叶礼堂	060 VANKE TRIPLE V GALLERY 万科三V画廊	106 TEA HOUSE 茶亭
012 ICD/ITE RESEARCHPAVILION 2011 斯图加特大学 ICD/ITKE 研究亭 2011	064 MAIN ENTRANCE GATE TO TIERRA CÁLIDA TIERRA CÁLIDA 的主入口	114 SERPENTINE GALLERY: THE RED SUN PAVILION 蛇形画廊：红色遮阳亭
018 ICD/ITE RESEARCHPAVILION 2010 斯图加特大学 ICD/ITKE 研究亭 2010	068 INCHON TRI-BOWL 仁川三碗	120 PARTY ANIMAL 派对动物
024 FIREPLACE FOR CHILDREN 儿童户外壁炉	074 CONGRESS CENTER HANGZHOU 杭州市政中心	126 RINGS AROUND A TREE 树之环
030 WOODS OF NET 千木巨网	078 SARPI BORDER CHECKPOINT 萨尔皮边境检查站	132 BURNHAM PAVILION 伯纳姆展亭
036 FRAGILE SHELTER 冬季避寒所	084 MARTIN LUTHER KIRCHE HAINBURG 马丁路德教堂	136 ON THE CORNER 街角住宅
044 VAGUE FORMATION VAGUE FORMATION 音乐亭	088 IGUZZINI IGUZZINI 照明公司总部	142 V-528 MULTIFUNCTIONAL ACTIVITY CENTER V-528 多功能活动中心
048 NY400 DUTCH PAVILION 纽约 400 荷兰馆	094 KUMUTOTO TOILET KUMUTOTO 卫生间	148 TIGER AND TURTLE - MAGIC MOUNTAIN 老虎与乌龟：魔山
052 ESKER HAUS 蛇形丘别墅	098 KIVIK ART PAVILIONS KIVIK 艺术展厅	152 FOOTBRIDGE IN EVRY 埃夫里人行桥
056 VENNESLA LIBRARY AND CULTURE HOUSE VENNESLA 图书馆和文化中心	102 TREEHOUSE DJUREN DJUREN 树屋	158 WOVEN BRIDGE 编织的桥

| 162 | ZAPALLAR PEDESTRIAN BRIDGE 札帕拉尔人行天桥 | 216 | COCOON 蚕茧大厦 | 268 | ROOFTECTURE HH ROOFTECTURE HH 私人住宅 |

168 RIGNY BRIDGE
RIGNY 桥

172 KURILPA BRIDGE
KURILPA 桥

180 SUNWELL MUSE
SUNWELL 缪斯

184 AEFAUP TEMPORARY BAR
AEFAUP 临时酒吧

188 UNIQLO CUBES
优衣库方块

194 SPORTS-PAVILION
鹿特丹体育俱乐部

198 HAPPY STREET
快乐街

202 BAMBOO RESTAURANT
竹子餐厅

208 747 WING HOUSE
747 机翼屋

220 GATE 750
大门 750

224 ISERNIA GOLF CLUB
伊塞尔尼亚高尔夫俱乐部

230 KOBAN
派出所岗亭

236 KALMAR MUSEUM OF ART
卡尔马艺术博物馆

242 MIRROR HOUSE
镜宅

250 NORWEGIAN WILD REINDEER CENTRE PAVILION
挪威野生驯鹿中心

254 YUSUHARA WOODEN BRIDGE MUSEUM
梼原町木桥博物馆

260 ONE OCEAN
2012 丽水世博会主题亭

264 ROCK IT SUDA
ROCK IT SUDA 度假屋

272 THE WELLINGTON ZOO HUB
惠灵顿动物园活动中心

276 AKRON ART MUSEUM
阿克伦城艺术博物馆

282 BMW WELT
宝马世界

288 ART GALLERY OF ALBERTA
加拿大阿尔伯塔省艺术博物馆

294 TROLLWALL RESTAURANT
TROLLWALL 饭店

300 BUSAN CINEMA CENTER
韩国釜山电影中心

306 WESTEND GATE
万豪酒店

312 RIZHAO LANDSCAPING PROJECT
日照山海天阳光海岸配套公建

LEAF CHAPEL
树叶礼堂
NIGHT WEDDING RESORT 夜间婚礼胜地

Location: Kobuchizawa, Yamanashi, Japan
Completion year: 2004
Designer: Klein Dytham Architecture
Use: Chapel
Site Area: 168 m²

项目地址：日本山梨县小渊泽町
竣工时间：2004 年
项目设计：日本 Klein Dytham 建筑事务所
建筑用途：小礼堂
项目面积：168 m²

Project Introduction

The Leaf chapel sits within the grounds of the Risonare hotel resort in Kobuchizawa, a refreshingly green setting with beautiful views to the southern Japanese Alps, Yatsugatuke peaks and Mt.Fuji.

项目介绍

该项目位于小渊泽的 Risonare 度假村。在这个清新迷人的建筑内，可以欣赏到日本南部阿尔卑斯山脉、Yatsugatuke 峰顶及富士山的美景。

Design Conception

The chapel is formed by 2 leaves - one glass, one steel - which have seemingly fluttered to the ground. The glass leaf with its delicate lace pattern motif emulates a pergola and the structure holding it up reminds one of the veins as of a leaf which slowly become thinner the further they get away from the central stem. The white steel leaf, perforated with 4,700 holes, each of which hold an acrylic lens, is similar to bride's veil made of delicate lace. Light filters through the lenses and projects a lace pattern onto the white fabric inside. Throughout the day as the sun turns the projected pattern naturally changes to create a myriad of different lace patterns on the inner lining of the veil creating a marvelous background for the wedding ceremony.

Layout Plan —— 规划图   Site Plan —— 平面图

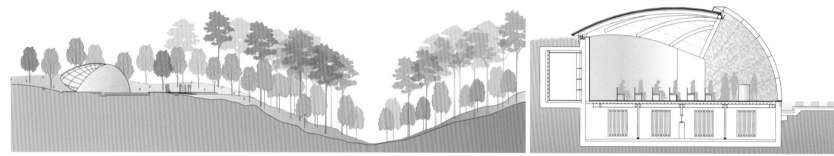

Sectional Drawing —— 剖面图

设计构思

教堂由两片叶子组成，一片是玻璃的，另一片是钢铁的。它们似乎刚刚从空中飘落下来。有着精致饰带的玻璃树叶，结构上像一个棚架，将屋面撑起。在渐渐远离主干的地方，叶面慢慢变薄，其上的纹理也与真实的叶子相仿。白色的钢结构树叶，穿有 4 700 个孔，每个孔都镶有一个丙烯酸有机玻璃透镜，它就像新娘面纱上的精致蕾丝。光线穿过透镜，将图案投射到建筑内部白色的织物上。随着太阳的移动，投射的图案不断变化，给婚庆典礼营造出梦幻般的场景。

ICD/ITKE RESEARCH PAVILION 2011
斯图加特大学 ICD/ITKE 研究亭 2011
BIONIC BUILDING 仿生建筑

Location: Stuttgart, Germany
Completion year: 2011
Designer: ICD/ITKE
Use: Research

项目地址：德国斯图加特
竣工时间：2011 年
项目设计：ICD/ITKE
建筑用途：研究

Project Introduction

In the summer of 2011 the Institute for Computational Design (ICD) and the Institute of Building Structures and Structural Design (ITKE), together with students at the University of Stuttgart have realized a temporary, bionic research pavilion made of wood at the intersection of teaching and researching.

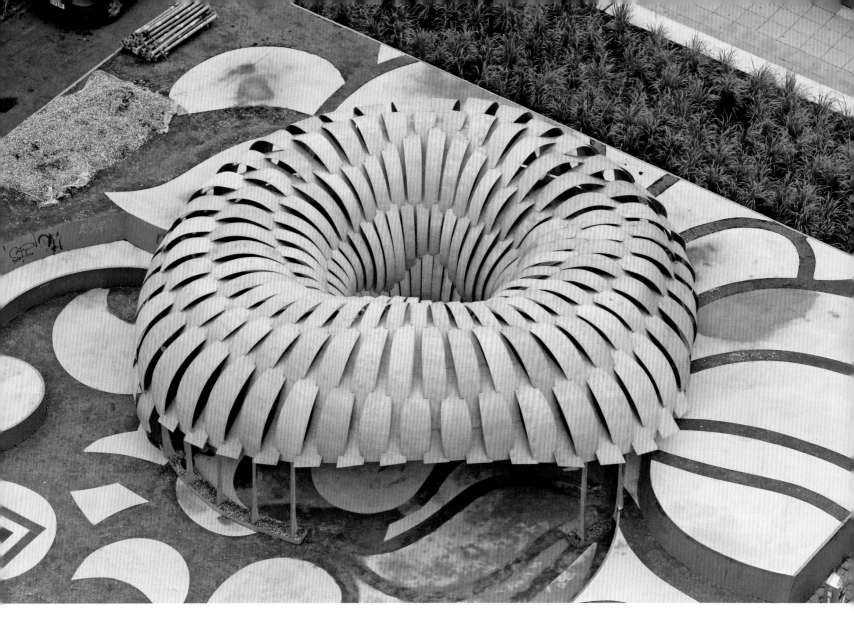

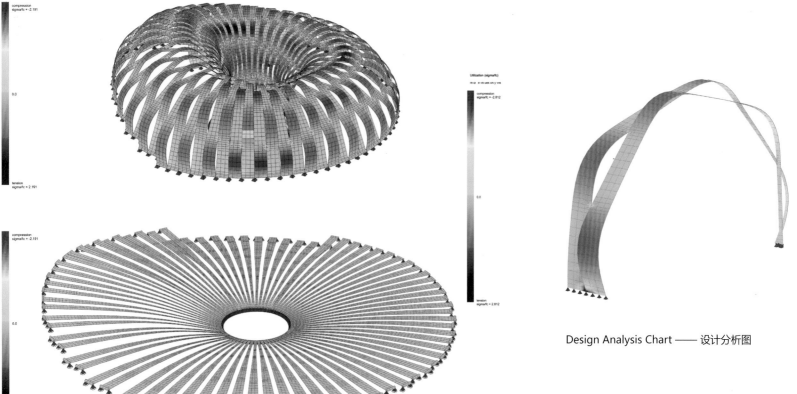

Design Analysis Chart —— 设计分析图

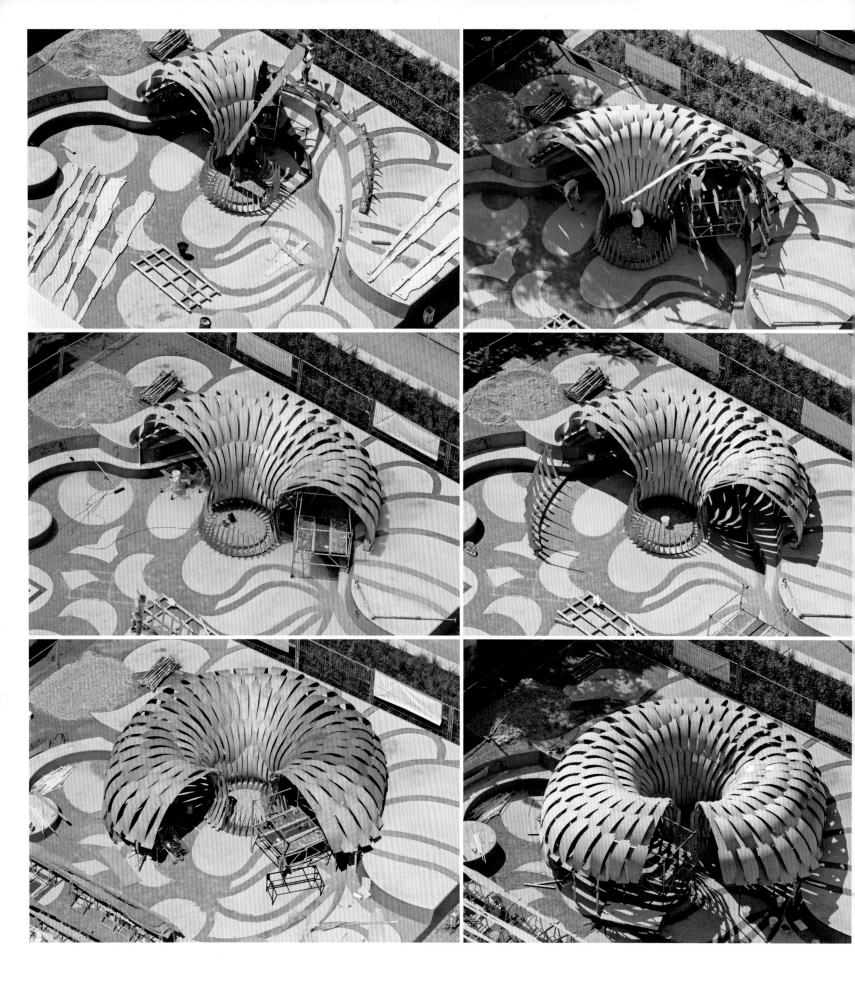

项目介绍

2011年夏，计算设计协会和建筑结构与结构设计协会联合斯图加特大学学生共同设计了一个临时性的仿生研究亭，亭子由纯木制造，用于教学和研究。

Design Conception

The project explores the architectural transfer of biological principles of the sea urchin's plate skeleton morphology by means of novel computer-based design and simulation methods, along with computer-controlled manufacturing methods for its building implementation. A particular innovation consists in the possibility of effectively extending the recognized bionic principles and related performance to a range of different geometries through computational processes, which is demonstrated by the fact that the complex morphology of the pavilion could be built exclusively with extremely thin sheets of plywood (6.5 mm).

设计构思

该项目通过新颖的计算机设计和仿真模拟法,以及在建筑中使用计算机控制制造法,将海胆的生物骨骼结构转变成一种建筑风格。这项极富特色的创新项目主要在于将可辨识的生物原理和相关的性能通过电脑程序高效地转变成不同的几何形状。这一过程可以证明,亭子形态只靠使用超薄的胶合板(厚6.5mm)就能建成。

Streifenpaare Teil 1

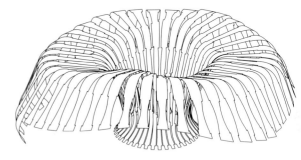

Streifenpaare Teil 2

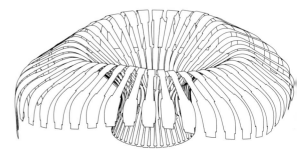

Spanten

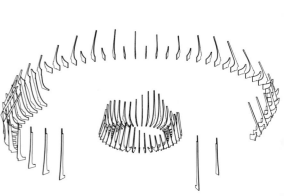

Kieskasten und Zugband

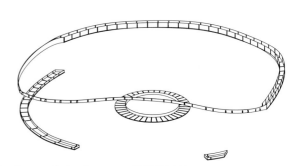

Design Analysis Chart —— 设计分析图

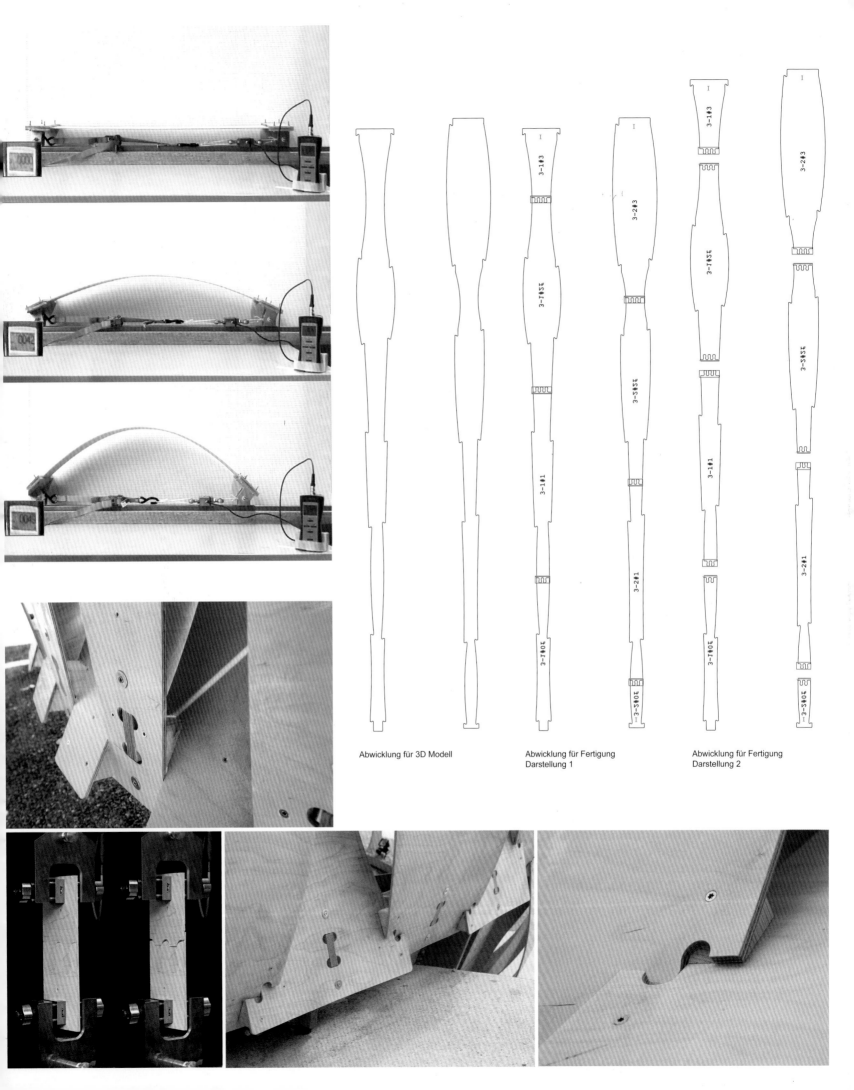

Abwicklung für 3D Modell

Abwicklung für Fertigung
Darstellung 1

Abwicklung für Fertigung
Darstellung 2

ICD/ITKE RESEARCH PAVILION 2010
斯图加特大学 ICD/ITKE 研究亭 2010
MAGIC MUSHROOM 魔力蘑菇

Location: Stuttgart, Germany
Completion year: 2010
Designer: ICD/ITKE
Use: Research

项目地址：德国斯图加特
竣工时间：2010 年
项目设计：ICD/ITKE
建筑用途：研究

Project Introduction

In July 2010, the Institute for Computational Design (ICD) and the Institute of Building Structures and Structural Design (ITKE), both of the University of Stuttgart, constructed a temporary research pavilion. The innovative structure demonstrates the latest developments in material-oriented computational design, simulation and production processes in architecture.

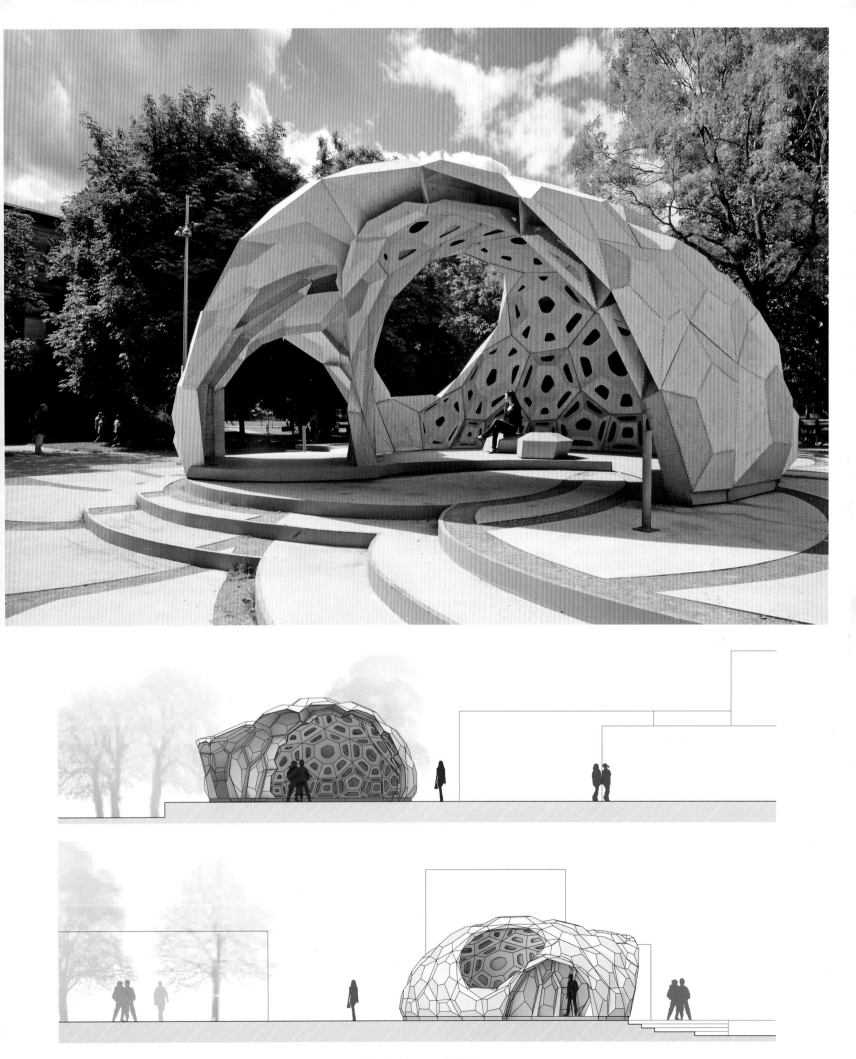

Rendering —— 效果图

项目介绍

2010年7月,斯图加特大学的计算设计协会和建筑结构与结构设计协会,共同设计了一个临时研究亭。这样创新的结构体现了材料导向电脑设计、仿真及建筑生产流程的最新技术。

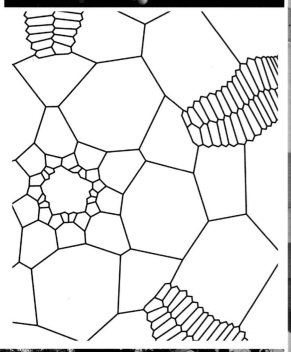

Design Conception

The building is a bending-active structure made entirely of elastically-bent plywood strips. The strips are robotically manufactured as planar elements, and subsequently connected so that elastically bent and tensioned regions alternate along their length. The force that is locally stored in each bent region of the strip, and maintained by the corresponding tensioned region of the neighboring strip, greatly increases the structural capacity of the system. In order to prevent local points of concentrated bending moments, the locations of the connection points between strips needs to change along the structure. The combination of both the stored energy resulting from the elastic bending during the construction process and the morphological differentiation of the joint locations enables a very lightweight system.

设计构思

这是一个完全由条状弯曲胶合板组成的主动挠曲结构。其错综复杂的网络结构体系由结点及相关的力矢量组成，利用胶合薄板的弹性加以调节，从而达到精妙的平衡状态，拓展出一方独特的建筑空间。初始的胶合平板条由六轴工业机器人制造并连接而成，且使其受弯和受拉部位横向交错分布。局部应力存储在板条的各个挠曲部位，并由相邻板条所对应的张拉部位加以维持，从而大大增加了系统的结构承载力。为了避免瞬间局部弯矩集中，板条之间的接头交叉排列，最终形成了新奇独特的建筑外壳。

FIREPLACE FOR CHILDREN
儿童户外壁炉
WARM FUN PLACE 温暖的游乐园

Location: Trondheim, Norway
Completion year: 2009
Designer: Haugen/Zohar Arkitekter
Use: Play
Client: Trondheim Municipality, Norway
Photographer: Jason Havneraas, Unni Skoglund & Grethe Fredriksen

项目地址：挪威特隆赫姆
竣工时间：2009 年
项目设计：Haugen/Zohar Arkitekter
建筑用途：游戏
开发商：挪威特隆赫姆市政
摄影师：Jason Havneraas, Unni Skoglund & Grethe Fredriksen

Project Introduction

It is an outdoor project for a kindergarten. Although the region is characterized by predominantly maritime climate and the weather varies considerably throughout the seasons, there is a popular saying "there is no such thing as bad weather only bad clothing", and all seasons are equally attractive to the Norwegian children that enjoy outside activities all year round.

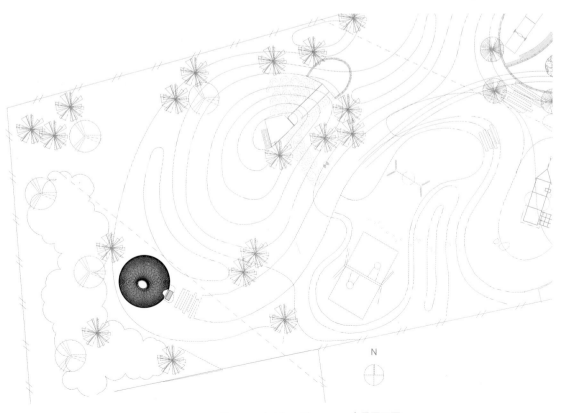

Panoramic View Plan —— 全景平面图

项目介绍

这是为幼儿园建造的一个户外项目。尽管当地属于典型的海洋性气候,但随着季节的交替,气候多变。当地有一种说法:"不是天气差,但就是容易穿错衣服。"然而对于孩子们来说,无论哪个季节,都不乏外出游玩的兴致。

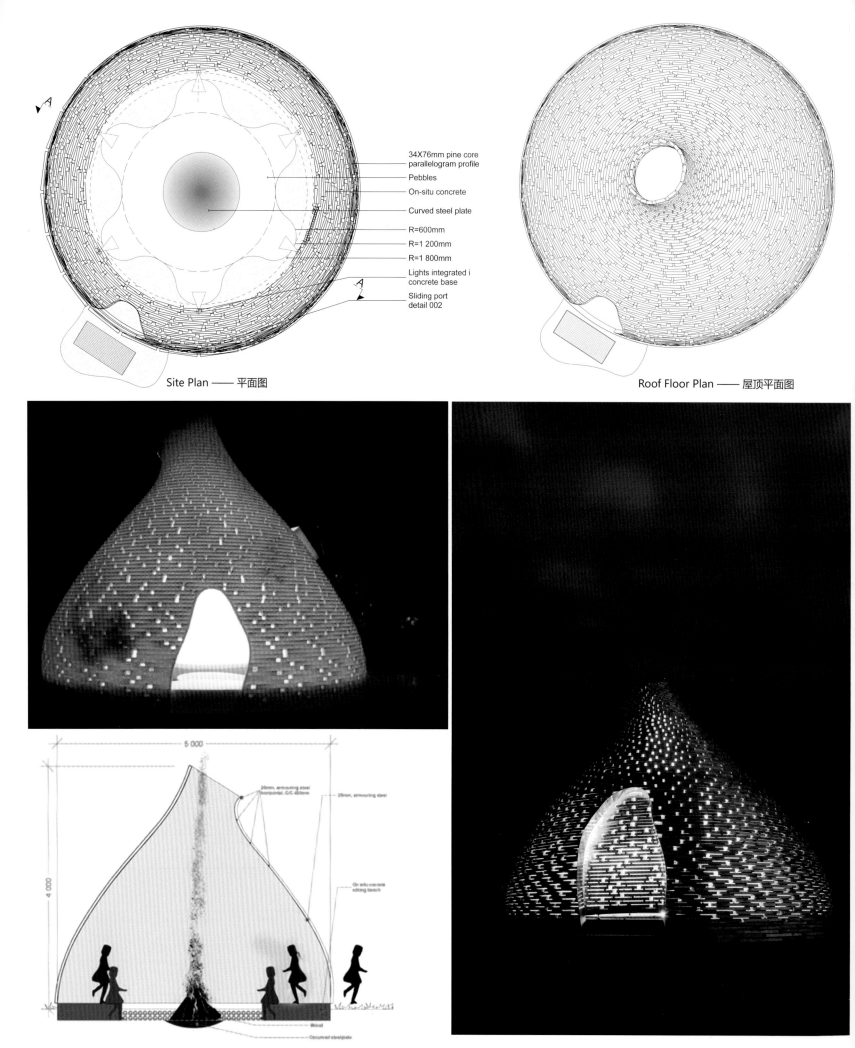

Site Plan —— 平面图

Roof Floor Plan —— 屋顶平面图

Sectional Drawing —— 剖面图

Design Conception

Together with the standard playground facilities the designer wished to create an enclosed space for fire, taletelling and playing. Given a very limited budget, reusing leftover materials (from a nearby construction site) was a starting point that led the design to be based on short wooden pieces. Inspired by the Norwegian turf huts and old log construction, a wooden construction was built and mounted on a lighted and brushed concrete base. The structure was made of 80-layered circles. Every circle was made of 28 pieces of naturally impregnated core of pine that are placed with varied spaces to assure chimney effect and natural light. A double curved sliding door was designed for locking the structure.

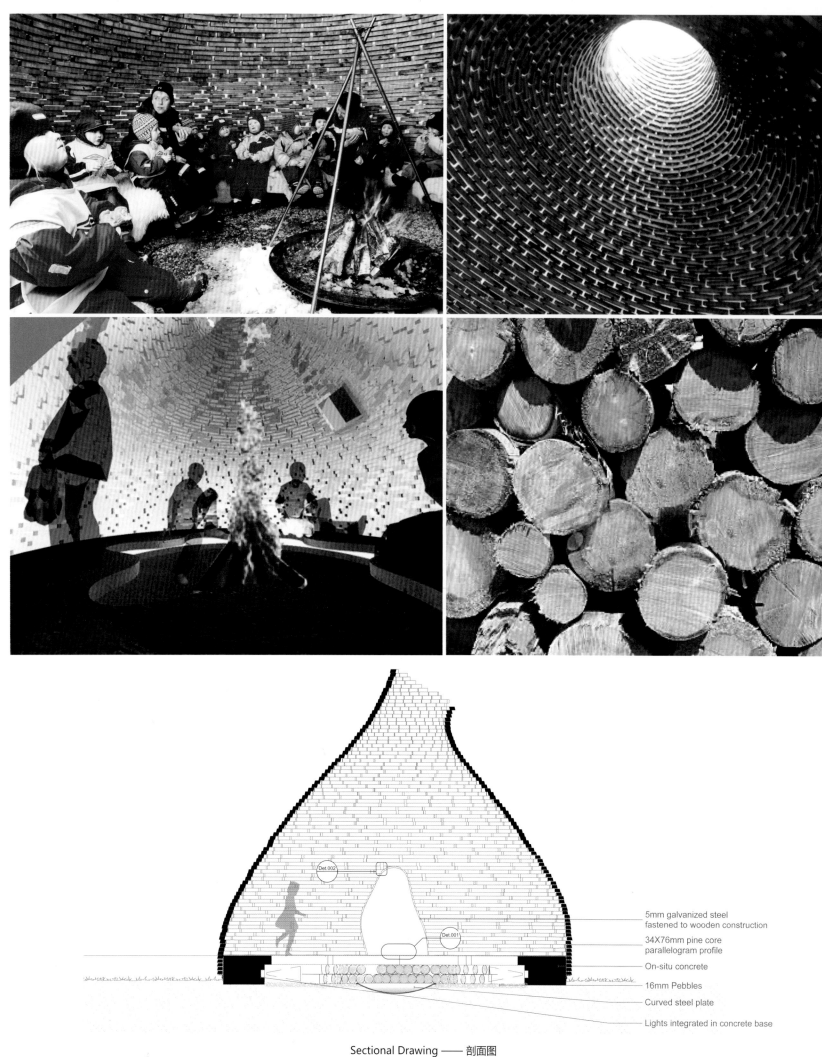

Sectional Drawing —— 剖面图

5mm galvanized steel fastened to wooden construction

34X76mm pine core parallelogram profile

On-situ concrete

16mm Pebbles

Curved steel plate

Lights integrated in concrete base

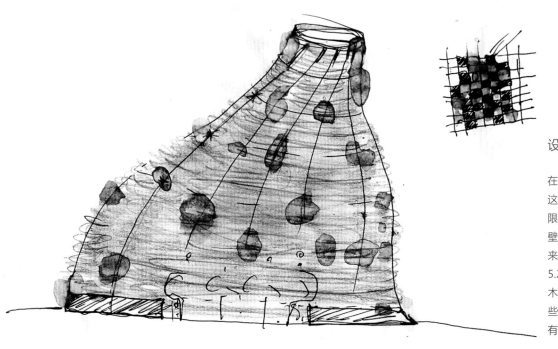

设计构思

在这个标准的操场上,设计师希望建造一个封闭的空间。在这里孩子们可以烤火、讲故事以及玩耍。因为预算非常有限,所以只能使用周围建筑工地的剩余材料来建造这个户外壁炉,因此建筑的设计以短木为基础。这个建筑的设计灵感来源于挪威的一个草坪小屋和旧式的原木建筑,它的尺寸为5.2m×4.5m。木制结构由80个圈组成,每个圈由28片松木构成。松木与松木之间形成了自然的缝隙,自然光透过这些缝隙进入到空间内部,保障了抽吸效应。此外,户外壁炉有一个双层的曲线形滑动门。

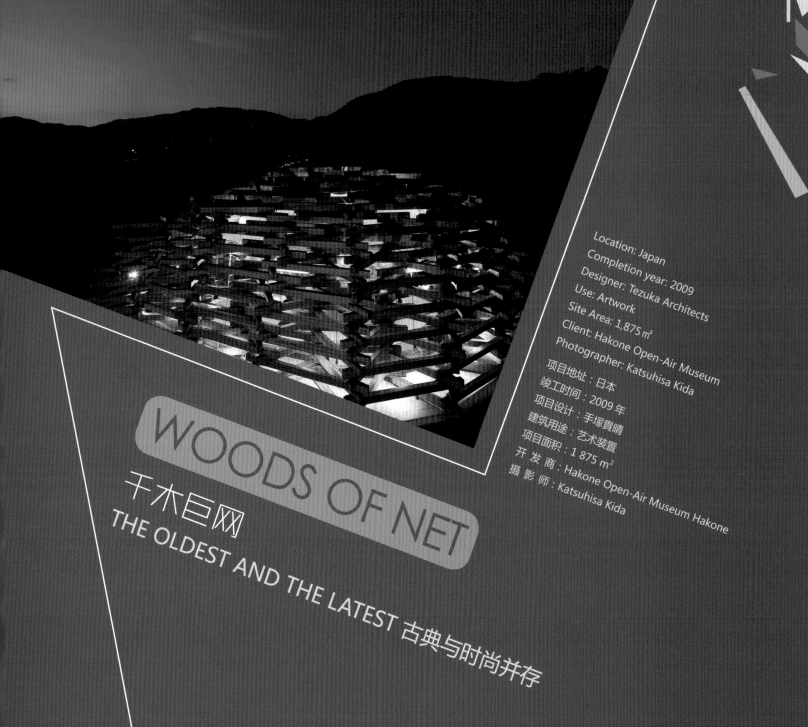

WOODS OF NET
千木巨网
THE OLDEST AND THE LATEST 古典与时尚并存

Location: Japan
Completion year: 2009
Designer: Tezuka Architects
Use: Artwork
Site Area: 1,875 ㎡
Client: Hakone Open-Air Museum
Photographer: Katsuhisa Kida

项目地址：日本
竣工时间：2009年
项目设计：手塚貴晴
建筑用途：艺术装置
项目面积：1 875 ㎡
开 发 商：Hakone Open-Air Museum Hakone
摄 影 师：Katsuhisa Kida

Project Introduction

This is a permanent pavilion for a net artist, Toshiko Horiuchi Macadam. The artist knitted the net entirely by hands, which is designed for children to crow in, roll around, and jump on the net. It was easy for us to see the artwork being outside even when it cannot be exposed to rain or ultraviolet light. We wanted to design a space as soft as the forest where the boundary between outside and inside disappears. The space attracts people like campfire. The children play inside the net just as fire and parents sit around or lay on the woods.

The structure is entirely composed of timbers without any metal parts. 320 cubic meter of timber members are used and there is nothing same among all the 589 members. The latest structural program was developed for the pavilion, but the joint techniques were derived from thousands years old Japanese wooden temples in Nara and Kyoto. As long as the proper maintenance is done, it is capable of existing over 300 years. This is the oldest and the most fashionoble structure in the world.

项目介绍

该项目是专门为织网艺术家 Toshiko Horiuchi Macadam 设计的，他用纯手工针织工艺完成了内部的编网，为孩子们提供了一个自由自在、爬上跳下的玩耍空间。手工织品虽不适合暴露在雨天和紫外线中，但在室外确实可以达到城市装饰的作用。于是，经过巧妙的设计，人们在户外也能欣赏五彩缤纷的编网。设计师希望打造出一个像森林一样安静祥和的空间，使巨网的内外不存在明显的界限。它就像个篝火晚会，吸引人们进来休憩、聚会。家长在周边的木墩上或躺或坐，看着孩子们兴致盎然地玩耍。

整座凉亭由 589 根大小不同、形状各异的木条建成，总体积达到了 320 m³，其中不含任何金属部件。在结构上，它采用了最先进的设计，然而其木材之间的链接咬合方式还是沿用着千年前奈良与京都建筑庙宇的技术。如能适当维护，巨网可以使用超过 300 年之久。它是世界上最具古风，同时也是最时尚的建筑。

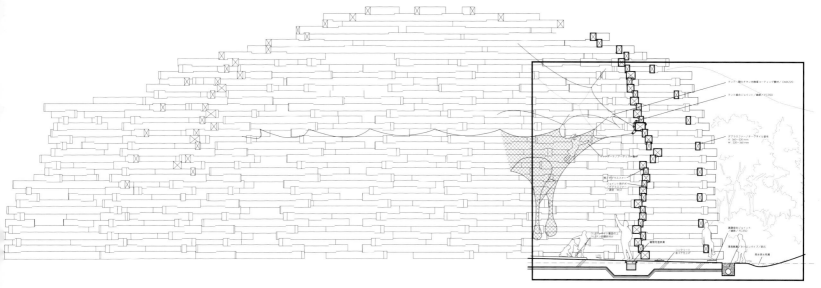

Structure Analysis Chart —— 结构分析图

SECTION DETAIL S=1:60

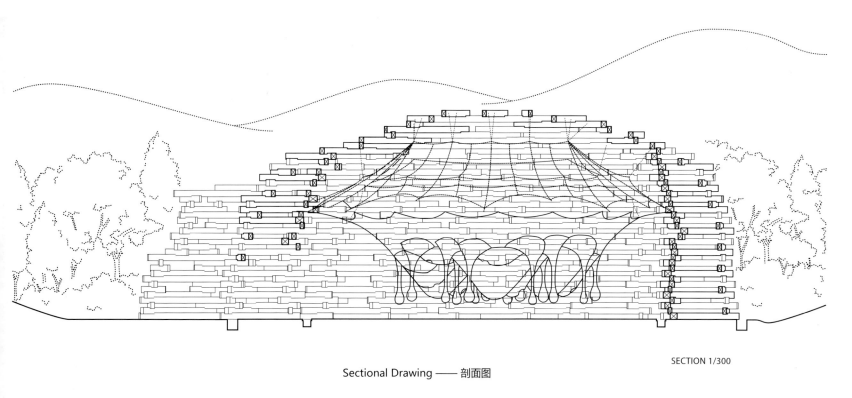

Sectional Drawing —— 剖面图

SECTION 1/300

Design Conception

The structural engineer, Professor Imagawa, developed a cutting-edge structural analysis program specifically for the pavilion. The program analyzed this new bending joint system overcome the variable characteristics of timbers. Through his analysis, he proved that the traditional technique was structurally sound. The members are solely connected by dowel pins and wedges, this Japanese traditional method could not be employed a few years ago under the strict Japanese building codes. Each member has a different size, In order to carry different kinds of loads from different directions. Each member has a complex shape so that water can easily drains to prevent the wood from warping.

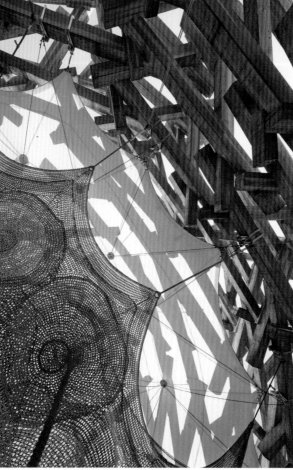

设计构思

建筑工程师金川教授特地为凉亭开发了尖端的结构分析程序，在分析弯曲关节系统时能够将木料的各种变化特征考虑在内。通过仔细分析，他最终决定使用传统的建筑技术。组成巨网的每一块木料部件各自都以板钉和木楔连接加固，严格来说这种日本传统建筑方法在几年前是不被允许的。之所以每块木料的大小都不相同，其实是为了能承受来自不同方向的作用力，而其复杂的结构有利于疏通积水以防木料变形。

FRAGILE SHELTER
冬季避寒所
SUBURB HUT 城郊小屋

Location: Hokkaido, Japan
Completion year: 2011
Designer: Hidemi Nishida
Use: Temporary Shelter
Photographer: Anna Nagai

项目地址：日本北海道
竣工时间：2011年
项目设计：Hidemi Nishida
建筑用途：临时避难所
摄 影 师：Anna Nagai

Project Introduction

Hidemi Nishida is a Sapporo-based artist, designer. Through Nishida's all projects, he has been trying to make a place that could make attentions to the surroundings. And he also tried to extract joyful happenings from the place. So Nishida's projects generally completed with gathering finally. This winter temporary shelter leads people together, having a party, staying, feeling the connection to the woods and so on in the white woods.

项目介绍

设计师 Hidemi Nishida 来自札幌，他一直在尝试让自己的作品与周边环境形成更为紧密的联系，并坚持不懈地从中挖掘快乐元素。因此，他的作品往往以提供公共活动空间为主要特色。于是我们可以看见，这间提供临时庇护的冬季避寒所，充分体现了 Hidemi Nishida 的设计理念。它让人们聚集在一起，休憩、聚会，进而与银装素裹的树林融为一体。

Design Conception

The shelter is made up of timber and wrapped in plastic sheeting. Although the latter is not the most sustainable material to use in any setting, it is durable and designed for reuse. This fragile function as being a shelter in the wild nature makes you a bit restless to stay there alone, but when few number of people get to gather into there, it's getting to be a wonderfully cozy, empathetic space. This is a reconfirmation of beginning of the "house". And it makes a primitive enjoyment of life.

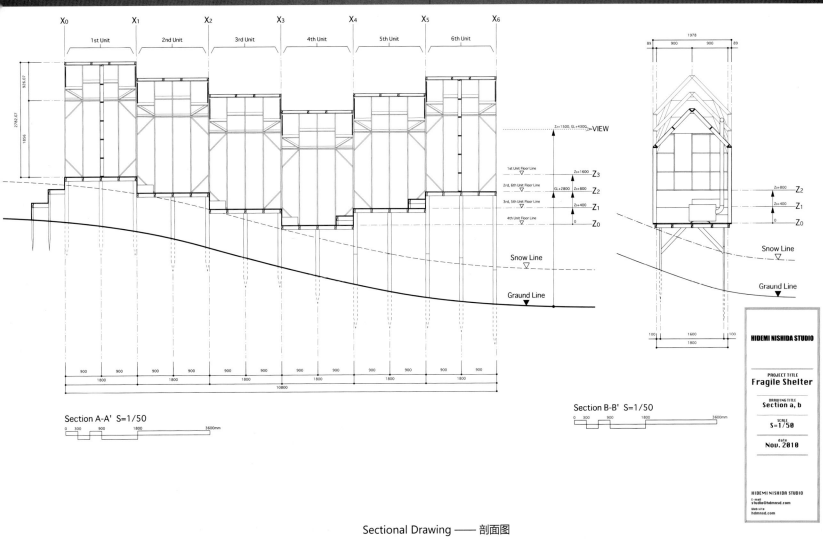

Sectional Drawing —— 剖面图

设计构思

避寒所主框架由木料搭成，外围则用半透明的塑料薄膜包裹着，基本达到防寒的要求。尽管塑料膜并不是最符合可持续性发展原则的材料，但它的好处在于使用时间较长，而且拆卸方便，可以重复使用。用表面看起来脆弱柔软的塑料膜来达到防寒目的，这不禁让身处荒野之地的人们感到一丝不安，尤其是独自一人时，内心的恐惧会愈演愈烈。但如果聚集而来的人越来越多，避寒所将会变得格外温暖、惬意，就像一个"家"。这时候，人们开始体验到一种极为朴素和简单的生活情趣。

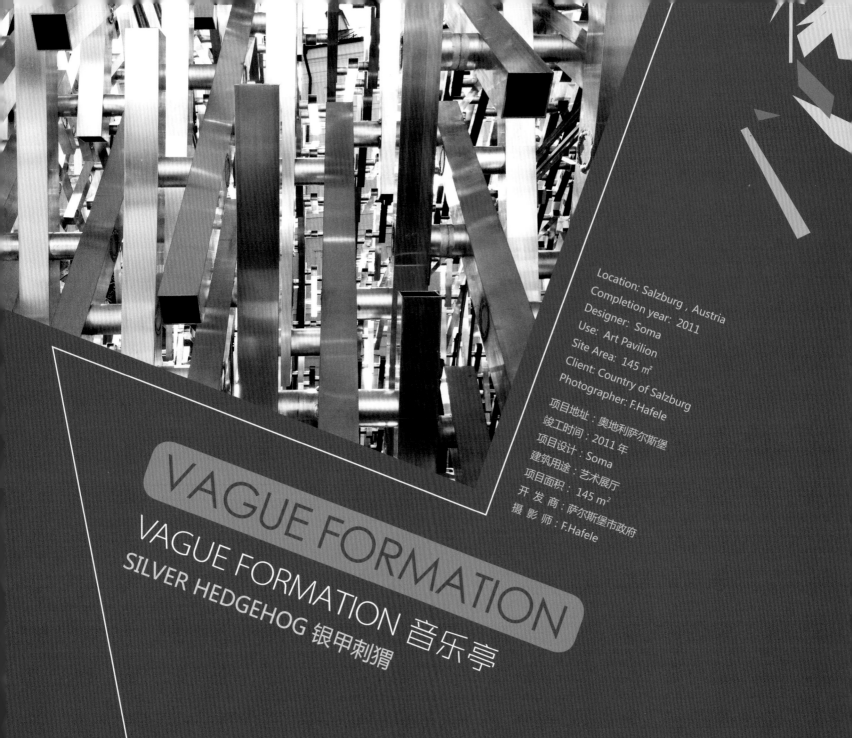

VAGUE FORMATION 音乐亭
SILVER HEDGEHOG 银甲刺猬

Location: Salzburg, Austria
Completion year: 2011
Designer: Soma
Use: Art Pavilion
Site Area: 145 m²
Client: Country of Salzburg
Photographer: F.Hafele

项目地址：奥地利萨尔斯堡
竣工时间：2011 年
建筑设计：Soma
建筑用途：艺术展厅
项目面积：145 m²
开 发 商：萨尔斯堡市政府
摄 影 师：F.Hafele

Project Introduction

The temporary pavilion created an unique presence for contemporary art productions in Salzburg: a city known predominantly for classical music. The main user of the pavilion is the Salzburg Biennale, a contemporary music festival. During the next decade it will be used for various art events in different cities.

项目介绍

这个临时的展厅为萨尔斯堡的现代艺术产品创造了一个独特的展示空间。众所周知，萨尔斯堡以古典音乐闻名世界。展厅设置的初衷，正是为萨尔斯堡双年展音乐节服务。随后的十年，它将会随着不同的艺术活动而周游各地。

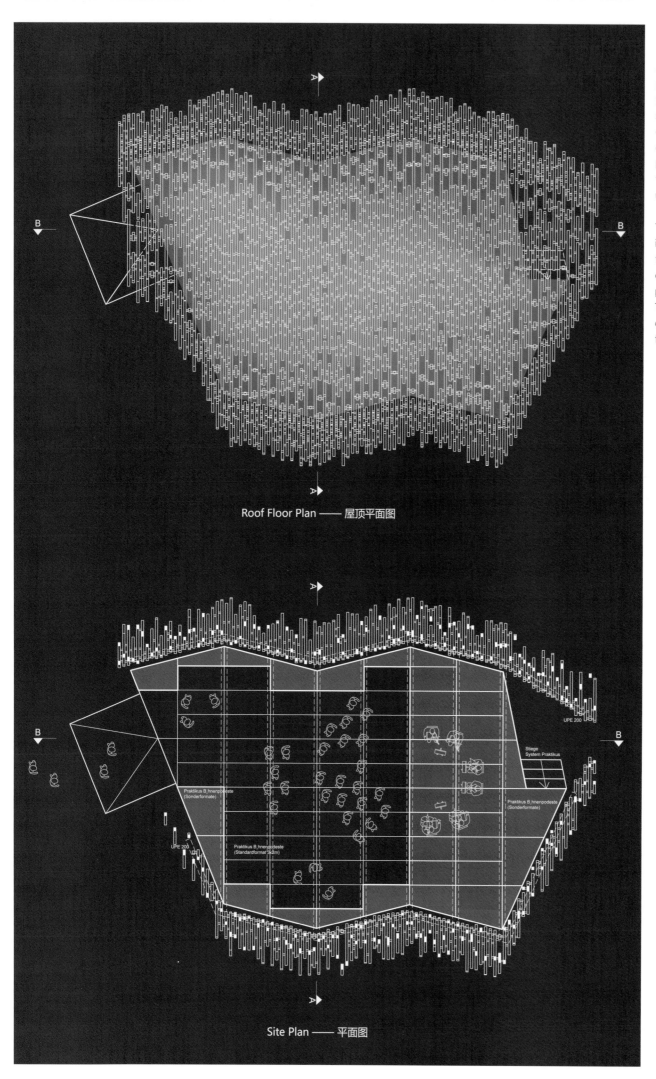

Roof Floor Plan —— 屋顶平面图

Site Plan —— 平面图

Design Conception

Art is a cultural process involving many participants within a discourse. This process does not present itself at first sight, but unfolds through encounter and engagement. The pavilion's appearance emphasises this idea. It provokes curiosity and invites visitors to encounter the unknown and unusual.

The structure can be divided into individual segments. By combining these segments in different ways or by reducing their number, the pavilion can adapt to its location. The removable interior membrane and the adjustable floor increase the flexibility of use.

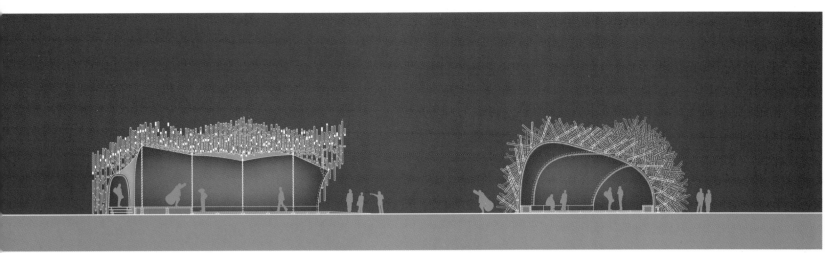

Sectional Drawing —— 剖面图

设计构思

艺术的形成是一个需要大众参与探讨的文化过程。它不会在第一眼便呈现全貌，需要经历从相遇到相知的过程。该项目的外观便突出了这样的理念。它以其未知和稀有，激发了人们的好奇心。从结构上来说，它可以分成若干独立的部件。这些部件通过不同形式的组装，或改变它们的数量，展厅可以适应不同的场地。可移动的内部薄膜和可调节的地板，更是增强了其灵活性。

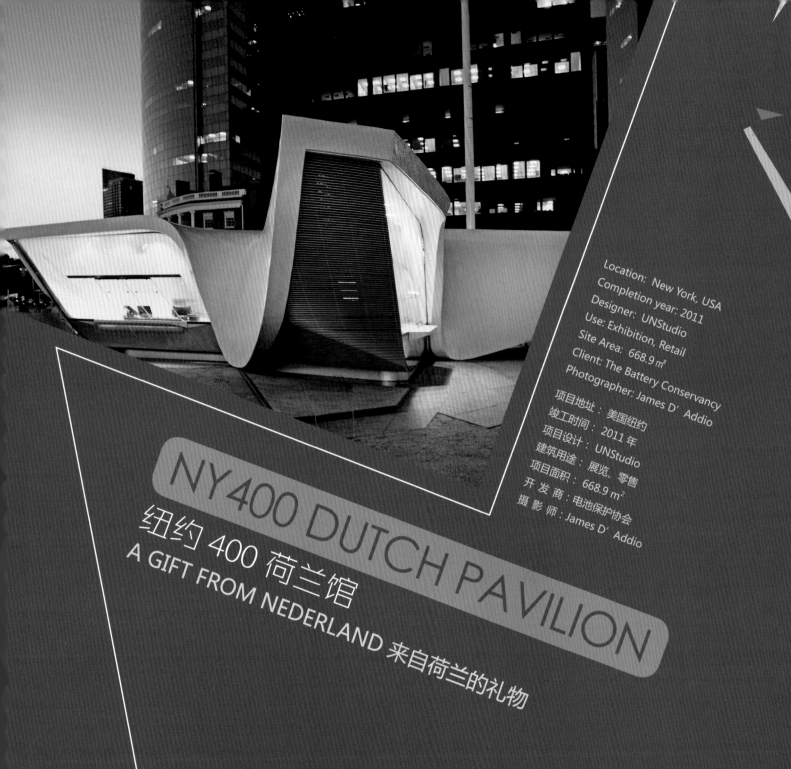

NY400 DUTCH PAVILION
纽约400荷兰馆
A GIFT FROM NEDERLAND 来自荷兰的礼物

Location: New York, USA
Completion year: 2011
Designer: UNStudio
Use: Exhibition, Retail
Site Area: 668.9 m²
Client: The Battery Conservancy
Photographer: James D' Addio

项目地址：美国纽约
竣工时间：2011年
项目设计：UNStudio
建筑用途：展览、零售
项目面积：668.9 m²
开 发 商：电池保护协会
摄 影 师：James D' Addio

Project Introduction

Placed on New Amsterdam Plein and commissioned by the Battery Conservancy, the NY400 Dutch Pavilion is presented as a gift from the Dutch government to the people of New York. The Pavilion is intended to introduce an opportunity for visitors, residents, and everyday commuters to pause and learn more at this historically important location. The Pavilion itself will be open with varying degrees throughout the day to the high number of commuters, tourists, and local residents.

项目介绍

该项目位于新阿姆斯特丹广场,是荷兰政府为了纪念荷兰与纽约400多年的友谊,委托电池保护协会建造的,是荷兰政府赠予纽约人民的一份礼物。它旨在为游客、当地居民和上班族提供一个休憩和学习的场所。由于每天来此的人数众多,所以,该馆会全天不同程度地对外开放。

Design Conception

The programme of this Pavilion oscillates between facility services (culinary outlet and information point), and a dynamic art, light, and media installation. The geometry of the Pavilion expresses its programmatic intentions, with the centre of the installation designed for more permanent, enclosed functions. Compawed to the enclosed nature of the core, the formal figure of the structure becomes increasingly more fluid and dispersed away from this centre as it opens onto the immediate landscape of the surroundings. The attendant 'flowering' or opening of the four wings of the Pavilion responds to varying orientations on the site as well as variety within the main programme. Within each wing, the contrast between the inside and the outside is blurred through the expression of continuous geometry. The geometric loop is introduced to virtually obscure the boundary between the ceiling, wall and floor and to promote integration of the built with the surrounding park.

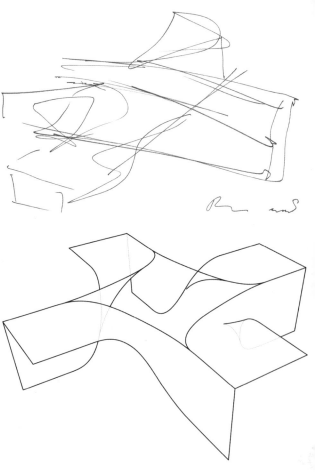

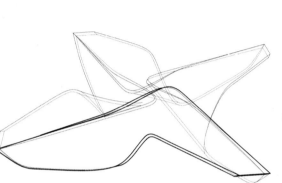
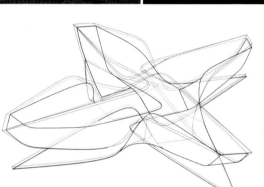
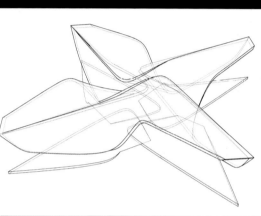

设计构思

该项目拥有双重功能——它既是一个服务设施（餐饮店和信息中心），又是一个极具活力的艺术、灯光和媒体装置。它的外形设计揭示了它的功能。建筑的中心地区是一个具持久性、封闭的功能区域。与中心区域相反，几个花瓣式的部分逐渐向周围景观扩散而去。而且越往外，每个"花瓣"就越具有流动感。每个"花瓣"及其四翼的开口（窗户）都朝着不同的方向，功能也各不相同。此外，几何弯曲的部分模糊了天花板、墙壁及地板之间的界限，使得建筑与周围公园的景观完美地融于一体。

ESKER HAUS
蛇形丘别墅
DECORATIVE PARASITE 装饰性子建筑

Location: Italy
Completion year: 2006
Designer: Plasma Studio
Use: Residence
Client: Patrick Holzer

项目地址：意大利
竣工时间：2006 年
项目设计：Plasma Studio
建筑用途：住宅
开 发 商：Patrick Holzer

Project Introduction

Esker Haus (esker=stratified geological formation) is a self-contained residential unit placed on the top of an existing house which was builded in the 1960s. The project has been developed as a parasite which started from adopting the structure of the host and gradually differentiated into its own unique organization and morphology.

项目介绍

该项目是一个独立的住宅单元,被加建在一座 20 世纪 60 年代老别墅的顶部。它就像一个寄生物,既延续了原建筑的结构,又逐步衍生出自己独特的组织和形态。

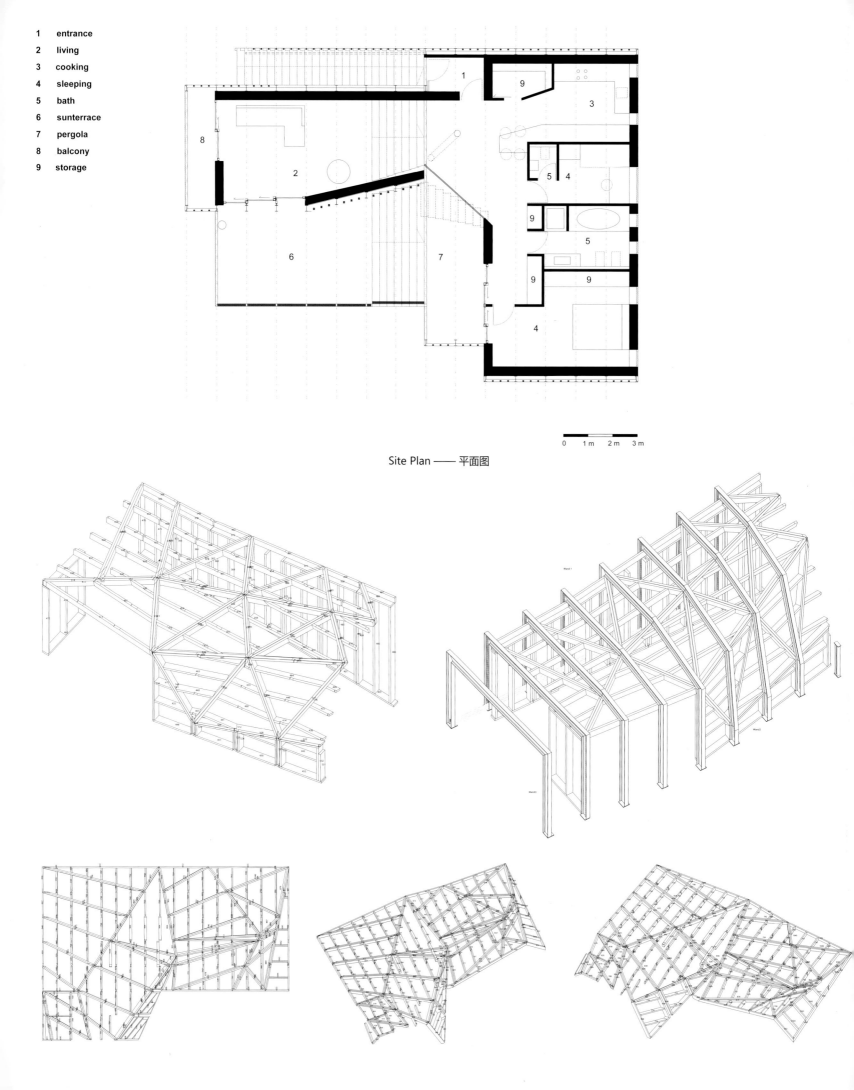

1	entrance
2	living
3	cooking
4	sleeping
5	bath
6	sunterrace
7	pergola
8	balcony
9	storage

Site Plan —— 平面图

Design Conception

The project is formed by a series of steel and timber frames that deform to recreate the smooth hillsides of the surrounding dolomites. The unique stratified morphology and construction system started off from projecting each step of the external staircase as a modulor that then was proliferated as frames. These frames enable the subsequent deformation and soften of the overall geometry.

设计构思

钢铁和框形支架组成的结构，映衬了周围白云石山脉光滑的轮廓。这个独特的分层式形态和结构系统从外部楼梯开始，设计团队把每一级台阶都当作精细的模型来设计，继而形成框架。这些框架实现了后来所有几何状物体的变形和软化。

VENNESLA LIBRARY AND CULTURE HOUSE
VENNESLA 图书馆和文化中心
INVITING PUBLIC SPACE 魅力空间

Location: Norway
Completion year: 2011
Designer: Vennesla Kommune
Use: Multipurpose
Site Area: 1,938 m²
Client: Vennesla Kommune
Photographer: Emile Ashley

项目地址：挪威
竣工时间：2011 年
项目设计：Vennesla Kommune
建筑用途：多功能
项目面积：1 938 m²
开 发 商：Vennesla Kommune
摄 影 师：Emile Ashley

Project Introduction

The new library in Vennesla contains a library, a cafe, meeting places and administrative areas, and links an existing residential area and learning centre together. Supporting the idea of an inviting public space, all main public functions have been gathered into one generous space allowing the structure combined with furniture and multiple spatial interfaces to be visible in the interior and from the exterior. Furthermore, the brief called for the new building to be open and easily accessible from the main city square, knitting together the existing urban fabric. This was achieved by using a large glass facade and urban loggia providing a protected outdoor seating area.

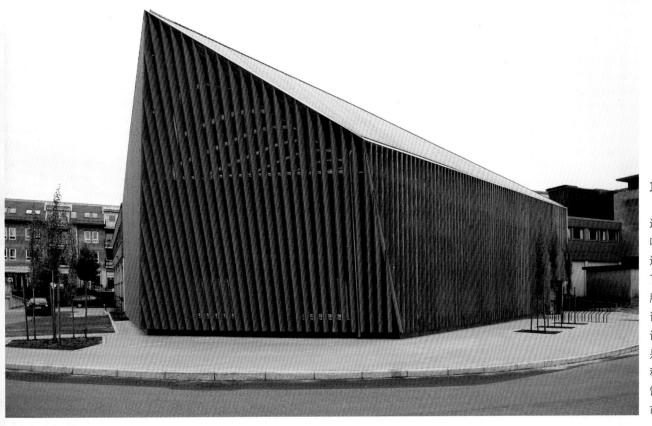

项目介绍

这个 Vennesla 的新图书馆包括图书馆、咖啡厅、会议室和办公区域等，它与附近的居民区和学习中心连接在一起。为了营造出一个富有魅力的公共空间，让所有的公共功能集聚在一个建筑内，设计师将家具功能附着于建筑结构上。无论在室内还是室外，不同空间的交接都是清晰可见的。此外，巨大的玻璃立面和户外休憩功能的凉亭走廊，让新图书馆变得更加开放，更好地融入原有的城市之中。

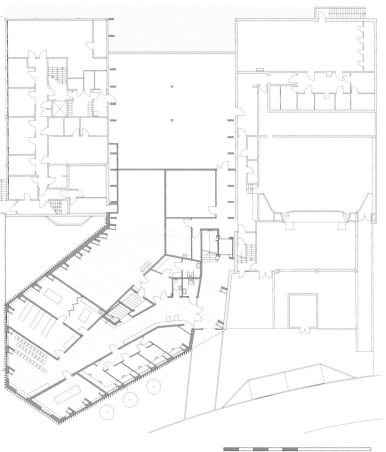

Site Plan —— 平面图

Design Conception

The designer deudopod a new rib shope building structure.In this project, we create an useable hybrid structures that combine a timber construction with all technical devices and the interior.A main intention has also been to reduce the energy need for all three buildings through the infill concept and the use of high standard energy saving solutions in all new parts. The library is a "low-energy" building, defined as class "A" in the Norwegian energy-use definition system. We aimed to maximize the use of wood in the building. In total, over 450 m³ of gluelam wood have been used for the construction alone. All ribs, inner and outer walls, elevator shaft, slabs, and partially roof, are all made in gluelam wood.

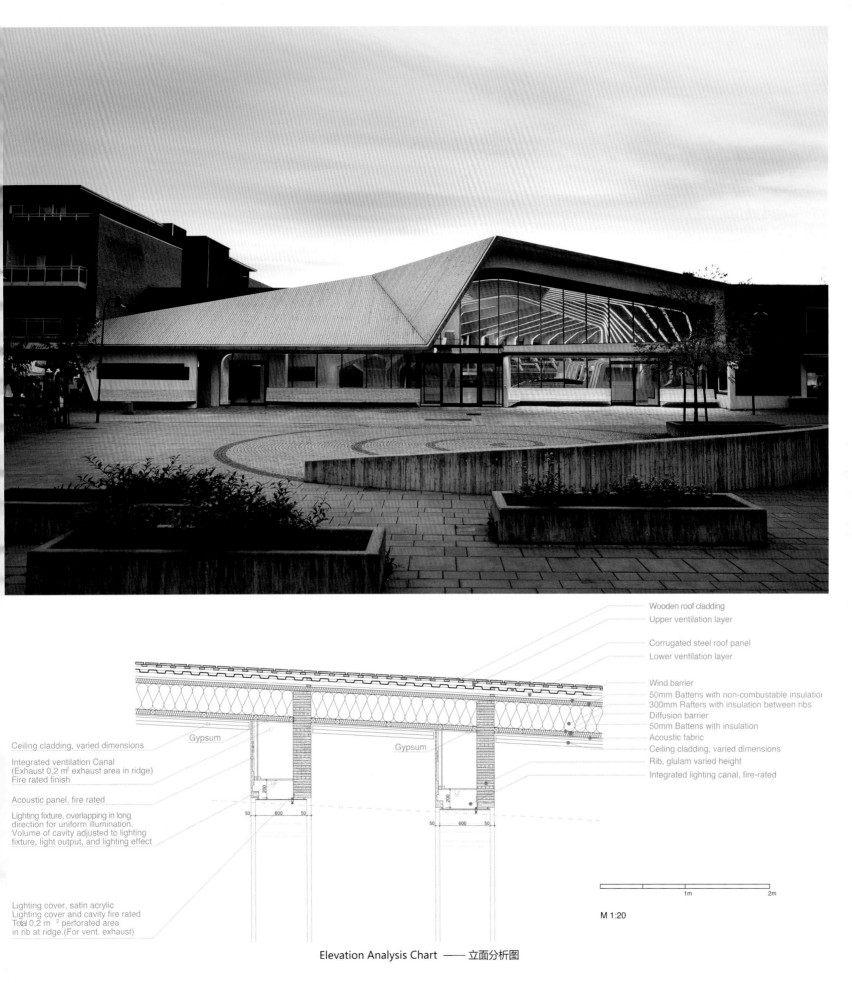

Elevation Analysis Chart —— 立面分析图

设计构思

设计师开发了一种新式肋状建筑结构。这个混合式木制结构涵盖了所有的技术设备和室内装饰。设计旨在减少建筑能耗，达到高标准的节能减排要求。换而言之，这个新图书馆是典型的低能耗建筑，达到了挪威能源系统评级中的 A 级低能耗环保建筑的标准。为了尽可能使用木质材料，设计师运用了 450 m³ 的胶合木板来完成该建筑，所有的支撑物、内外墙，甚至电梯、地板及部分的天花板都由胶合木版制作而成。

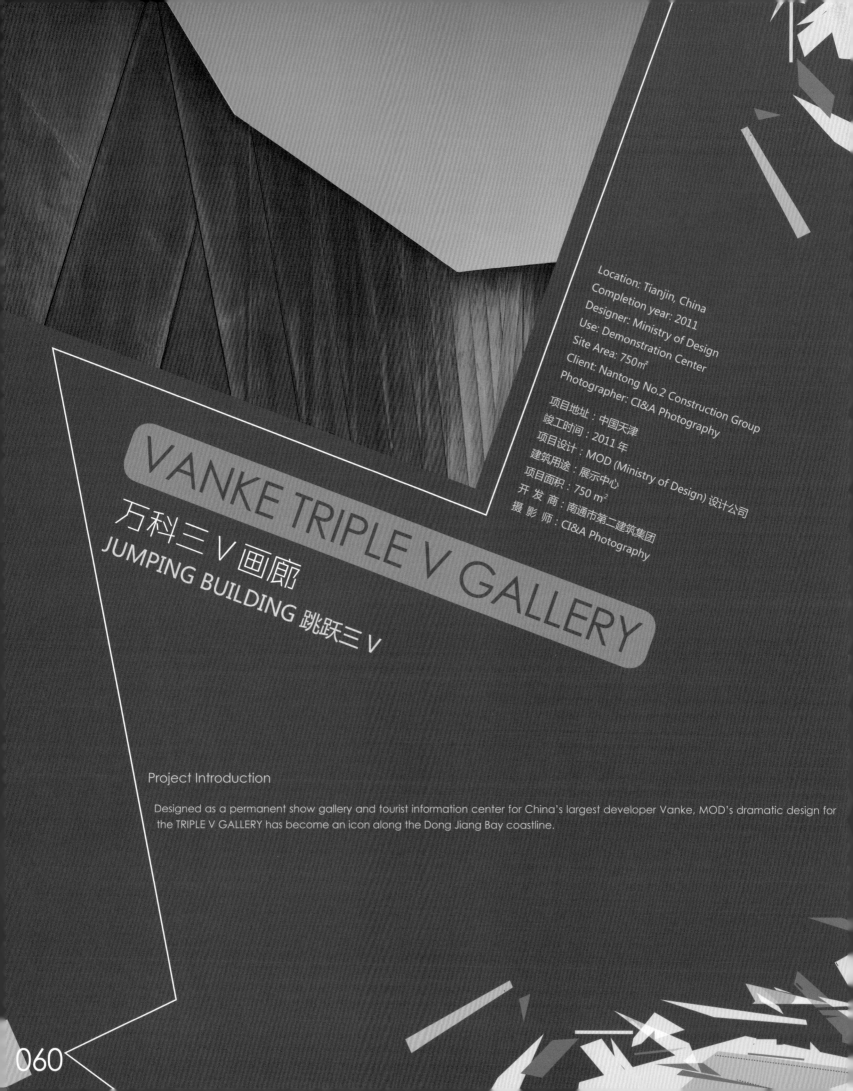

VANKE TRIPLE V GALLERY
万科三V画廊
JUMPING BUILDING 跳跃三V

Location: Tianjin, China
Completion year: 2011
Designer: Ministry of Design
Use: Demonstration Center
Site Area: 750m²
Client: Nantong No.2 Construction Group
Photographer: CI&A Photography

项目地址：中国天津
竣工时间：2011 年
项目设计：MOD (Ministry of Design) 设计公司
建筑用途：展示中心
项目面积：750 m²
开 发 商：南通市第二建筑集团
摄 影 师：CI&A Photography

Project Introduction

Designed as a permanent show gallery and tourist information center for China's largest developer Vanke, MOD's dramatic design for the TRIPLE V GALLERY has become an icon along the Dong Jiang Bay coastline.

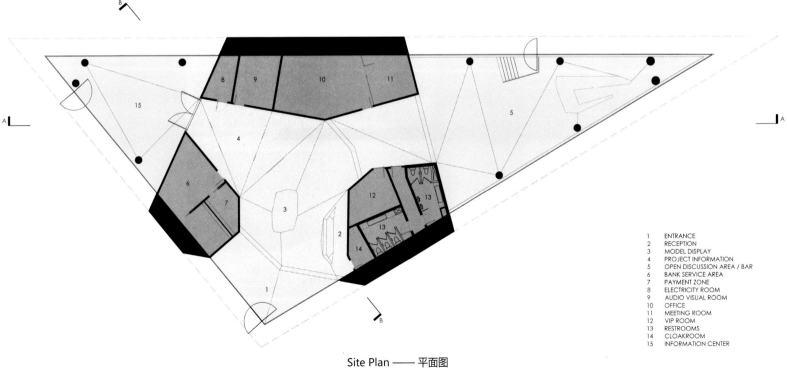

Site Plan —— 平面图

1. ENTRANCE
2. RECEPTION
3. MODEL DISPLAY
4. PROJECT INFORMATION
5. OPEN DISCUSSION AREA / BAR
6. BANK SERVICE AREA
7. PAYMENT ZONE
8. ELECTRICITY ROOM
9. AUDIO VISUAL ROOM
10. OFFICE
11. MEETING ROOM
12. VIP ROOM
13. RESTROOMS
14. CLOAKROOM
15. INFORMATION CENTER

项目介绍

设计团队为中国最大的房地产开发商万科集团设计了一个集永久性画廊和旅游信息中心于一身的项目。令人瞩目的造型，使它成为了东江湾海岸线上的标志性建筑。

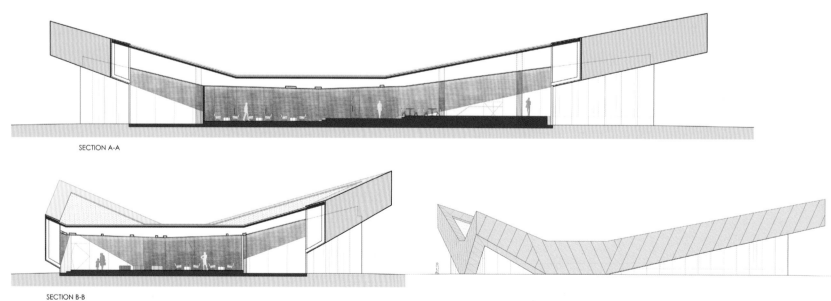

SECTION A-A

SECTION B-B

Design Conception

The client's program called for 3 main spaces: a tourist information center, a show gallery and a conferecehall. Despite its obvious sculptural qualities, the building's DNA evolved rationally from a careful analysis of key contextual and programmatic perimeters – resulting in the TRIPLE V GALLERY's triangulated floor plan as well as the 3 soaring edges that have come to define its form. Tectonically, the building responds to the coastal setting and is finished in weather-sensitive Corten steel panels on its exterior and timber strips on the interior walls and ceiling for a more natural feel.

Elevation Drawing —— 立面图

设计构思

该项目主要包括三个空间——旅游信息中心、画廊及一个会议厅。建筑具备明显的雕塑艺术特性，实际是通过对周围主要环境和规划区域的仔细分析后决定的。无论是三角形的地块还是耸起的三个角，都经过了细致的数据分析。该项目不仅从外形上呼应了沿海环境，材料的选择也是根据沿海气候而定。其室外的材料用到了耐腐蚀的考顿钢护板，室内用到了木板墙和木板天花以追求自然的感觉。

MAIN ENTRANCE GATE TO TIERRA CÁLIDA
TIERRA CÁLIDA 的主入口
A SILHOUETTE WITH WAVES 起伏的剪影

Location: Spain
Completion year: 2010
Designer: Manuel Clavel Rojo
Use: Gate
Site Area: 305 m²
Client: Grupo Urbanif S.L.
Photographer: David Frutos Ruiz

项目地址：西班牙
竣工时间：2010 年
项目设计：Manuel Clavel Rojo
建筑用途：大门
项目面积：305 m²
开 发 商：Grupo Urbanif S.L.
摄 影 师：David Frutos Ruiz

Project Introduction

This project was a result of restricted tender, as the main entrance gate of this architectural complex. And it series rnore thom 3,500 resodemts.

项目介绍

该项目的授权来自限制投标。作为建筑群的主入口,它所服务的住户超过 3 500 户。

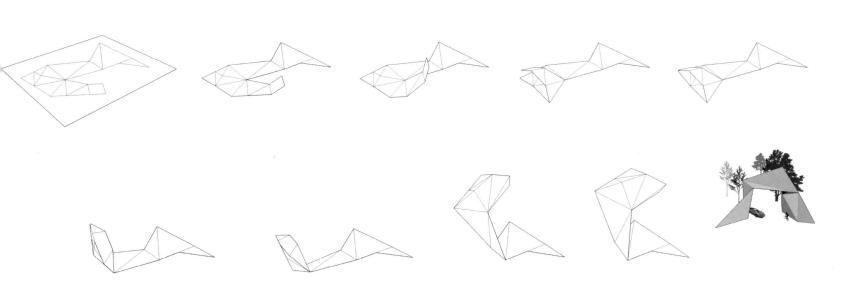

Design Analysis Chart —— 设计分析图

Design Conception

The project required that the gateway be made of just one structure in which the security control post would be housed and we placed in the fold of the construction. The lighting was installed in holes cut out of the concrete, thus integrating the illumination, converting the structure into a eye-catching vision, that changes from daytime, when it is almost organic and alive in texture and color, to evening when it takes on a more technological presence—a silhouette with waves of light that emphasize the entrance.

Rendering —— 效果图

 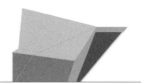 

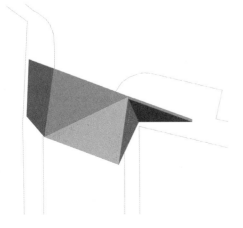 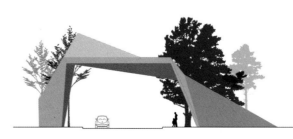

设计构思

大门只有一个框架结构。设计团队将安保点放在了室内，即褶层中。照明设备被安装在了孔洞之中。到了夜晚，灯光如星光，熠熠生辉，别具吸引力。这和白天以质感与色彩展现的美感不同，这时展现得更多的是科技感。灯光突出了入口起伏的轮廓。

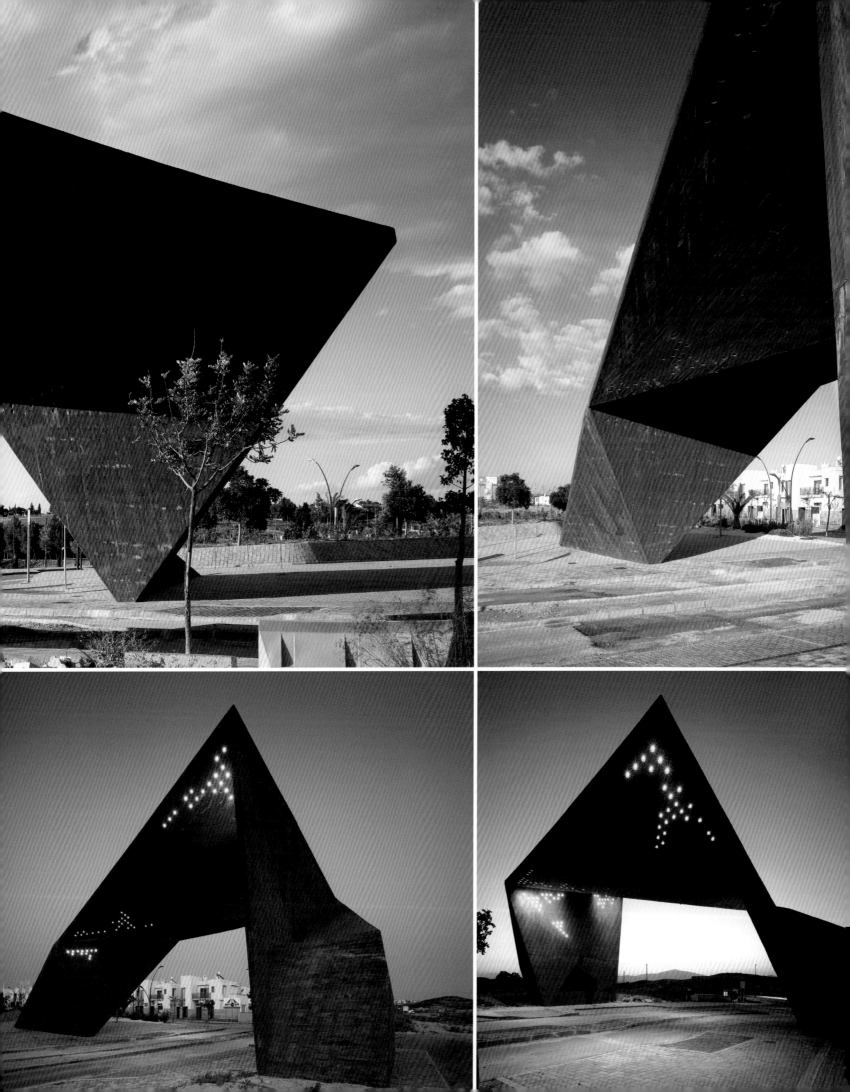

INCHON TRI-BOWL
仁川三碗
COMPOSITE CULTURAL SPACE 复合文化空间

Location: Incheon, South Korea
Completion year: 2010
Designer: iArc Architects
Use: Gallery
Site Area: 2,863.4 m²
Client: Young-chea Park

项目地址：韩国仁川
竣工时间：2010 年
项目设计：iArc Architects
建筑用途：画廊
项目面积：2 863.4 m²
开 发 商：Young-chea Park

Project Introduction

Incheon Tri-bowl is an uncommonly seen project that reverses the common understanding of an architectural space. Unlike the accustomed architecture space that applies small differences between the flat floor and ceiling, the Tri-bowl creates a free-curved floor under a flat ceiling.

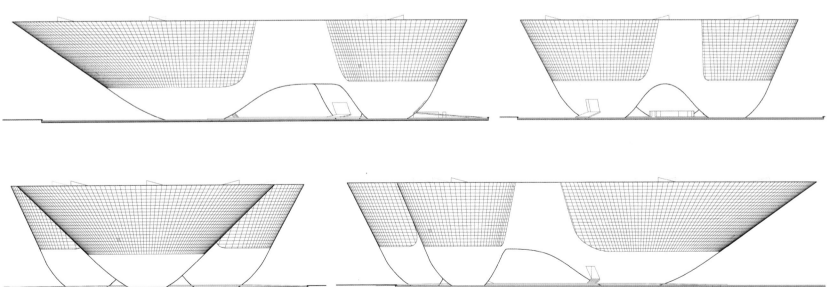

Elevation Drawing —— 立面图

项目介绍

该项目一改以往我们对建筑的认识，扩大了地板和天花板的差异，在平坦的天花板下安置了一个自由弯曲的地板。

Key-Plan

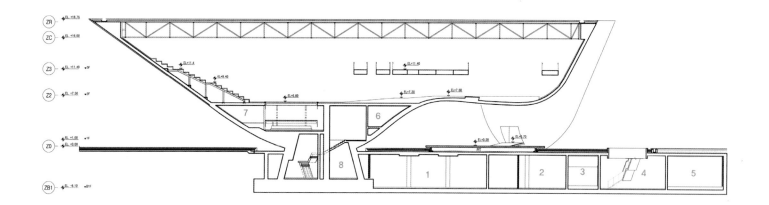

1. MACHANICAL ROOM
2. WATER TREATMENT ROOM
3. PROTECTION CONTROL ROOM
4. SUNKEN
5. ELECTRIC ROOM
6. WATING ROOM
7. STORAGE 1
8. STORAGE 2

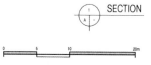

Sectional Drawing —— 剖面图

Design Conception

This structure floats on a reflecting pond where there is a long bridge under the extreme structure and the visitors can enter this structure through it. The bridge continues inside the building and acts as the main circulation of the building. The building is used as a gallery space. The programmatic space consists of an exhibition space, performance space and relaxation space. The theater can accommodate up to 400 people.

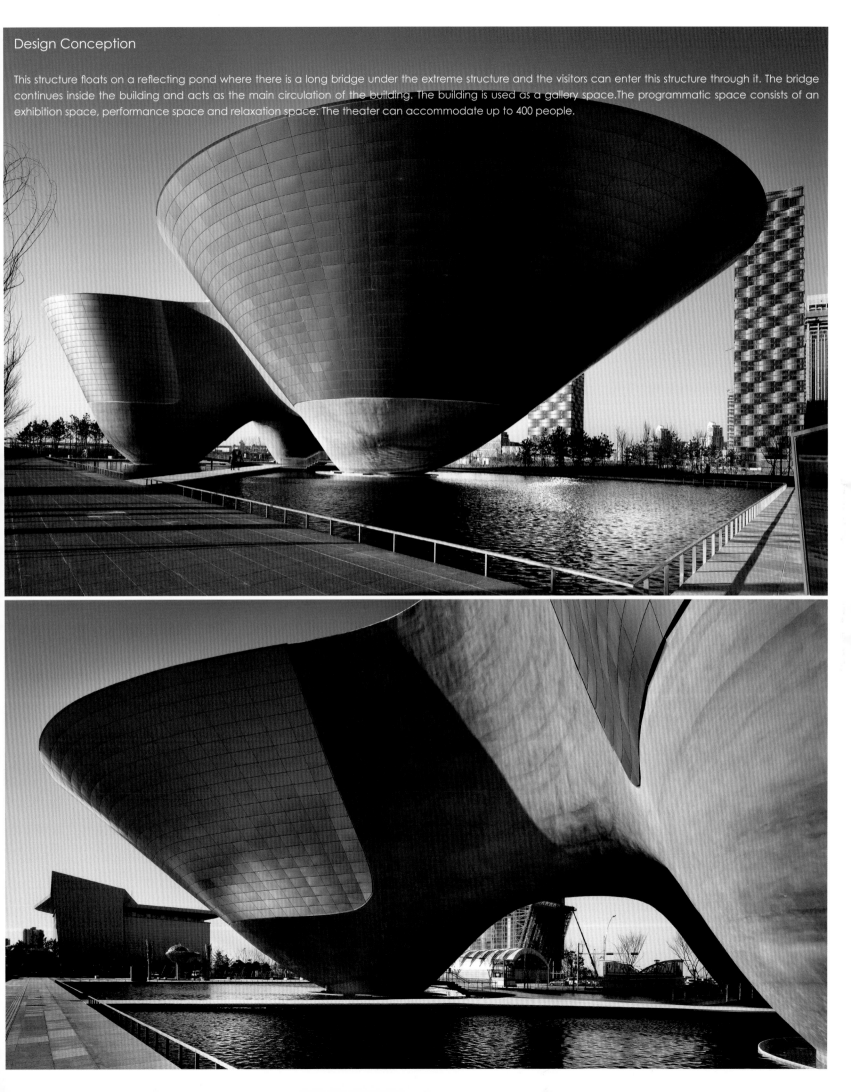

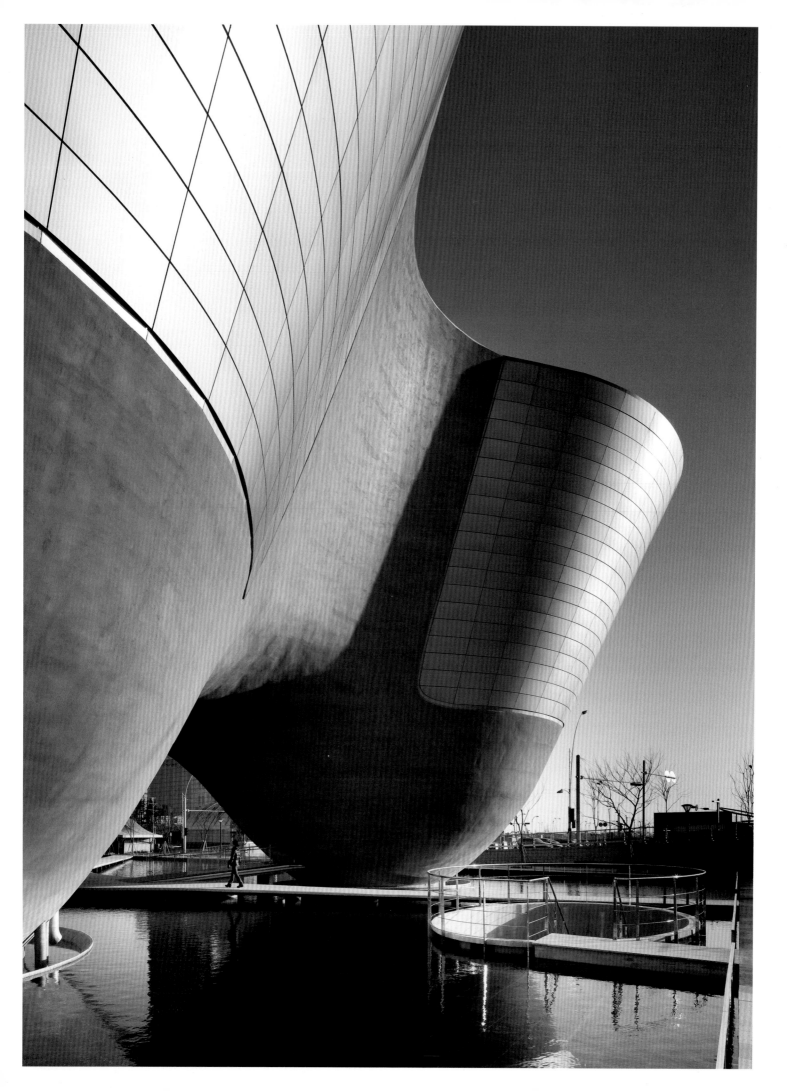

设计构思

这个独特的结构浮于水塘之上，建筑与湖面相映成趣。游客可以从其下的长桥进入画廊。桥梁一直延伸到室内，是该建筑的主要通道。这个画廊为游客提供了展览、表演和聚会的空间。内部的剧场和展览空间最多可以同时容纳 400 人参观。

CONGRESS CENTER HANGZHOU
杭州市政中心
BLOOMING LOTUS 盛开的莲花

Location: Hangzhou, China
Completion year: 2010
Designer: Peter Ruge Architekten
Use: Government Office
Site Area: 22,000 m²
Client: City of Hangzhou
Photographer: Jan Siefke

项目地址：中国杭州
竣工时间：2010 年
项目设计：Peter Ruge Architekten
建筑用途：政府办公
项目面积：22 000 m²
开 发 商：杭州市政府
摄 影 师：Jan Siefke

Project Introduction

The new building ensemble is situated close to Qiantang River and it is not far from the city centre. It will be the focus building of the new large business and administration district of the city. The new fascinating complex consists of six office high-rise buildings arranged in a circle and connects in the upper floors through a circular bridge building. The high-rise buildings are flanked with flat multi-functional buildings including four main entrances from all directions. As the new central form of the main administration building of the City of Hangzhou the Congress Centre resembles a large precious stone.

项目介绍

该项目位于钱塘江靠近市中心不远处,整体造型犹如一朵盛开的莲花。新的市政大楼建筑群由六座独立的办公大楼围成一个圆圈,并在每个办公楼的顶部之间通过一个环形的长廊彼此相连。整座建筑群有四个主要出入口,在建筑群正中心是会议中心,会议中心犹如宝石一样镶嵌在建筑群中间。

01 Top light
02 Roof garden
03 Steel construction (lamellas)
04 Facade construction

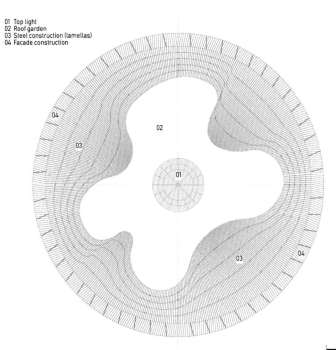

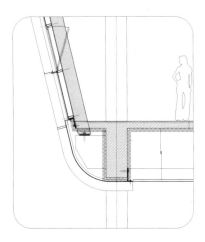

Design Conception

Zhejiang Province is known for its tea-producing region. To express the building's regional characteristics, design of the facade is based on the superimposed configurations of the tea cultivation pathways and the planting nets. As a result, the building is enveloped by a multi-layered fabric, giving it a true architectural plasticity. Looking from a distance, the facade appears like a rigid volume, but dissolves into a network of structures and levels as you come closer. The main idea for the design of the roof was to use it as the fifth facade of the building to set up a strong and typical local image in the shape of a lotus blossom, which you can see from all upper floors of the surrounding high-rise buildings. The facade structure would be extended unto the roof of the congress center to cover up it partly. Through the different lengths and fixed height of the steel beams the structure is waved and forms the abstract blossom of lotus in the center of the roof. This part which isn't covered is designed and planted as a green garden.

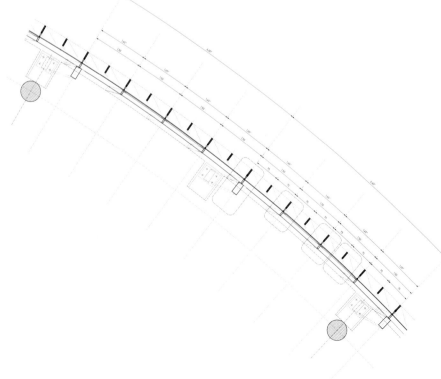

Detail - horizontal section
Scale 1:20

设计构思

浙江最有名的就是茶，为了表达建筑的地区性特征，表皮设计是以重叠交错的茶道和网格为基础的，最后建筑呈现出一个多层次的纹理，营造出一种真实的弹性感。从远处看，建筑表皮是不变的风格，但是走进了它就是一系列结构和层次的网络叠加。设计师将屋顶作为建筑的第五立面，从而反映当地的莲花文化。周围高层的表皮结构还能延伸以覆盖部分中间的会议中心。这个延伸的结构有不同的长度和固定的钢铁梁高度，它们互相交织形成抽象的莲花效果，没有覆盖的那部分则是一座绿色的楼顶花园。

SARPI BORDER CHECKPOINT
萨尔皮边境检查站
GREETING & VIEWING 迎宾瞭望台

Location: Sarpi, Georgia
Completion year: 2011
Designer: J. MAYER H. Architects
Use: Border Checkpoint
Site Area: 16,500 m²
Client: Ministry of Finance of Georgia
Photographer: Jesko Malkolm Johnsson-Zahn, Beka Pkhakadze

项目地址：格鲁吉亚萨尔皮
竣工时间：2011 年
项目设计：J. MAYER H. 建筑事务所
建筑用途：边境检查站
项目面积：16 500 m²
开 发 商：格鲁吉亚财政部
摄 影 师：Jesko Malkolm Johnsson-Zahn, Beka Pkhakadze

Project Introduction

The customs checkpoint is situated at the Georgian border to Turkey, besides the shore of the Black Sea.

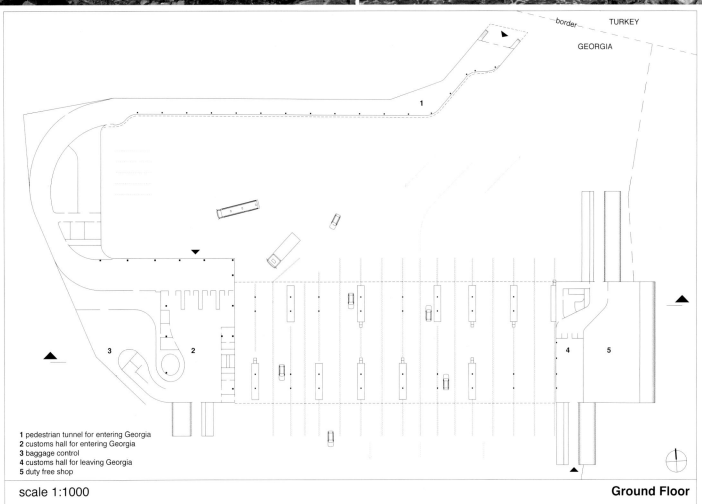

1 pedestrian tunnel for entering Georgia
2 customs hall for entering Georgia
3 baggage control
4 customs hall for leaving Georgia
5 duty free shop

scale 1:1000 Ground Floor

Panoramic View Plan —— 全景平面图

项目介绍

该项目位于格鲁吉亚和土耳其的边界——黑海海岸。

Design Conception

With its cantilevering terraces, the tower is used as a viewing platform, with multiple levels overlooking the water and the steep part of the coastline. In addition to the regular customs facilities, the structure also contain a cafeteria, staff rooms and a conference room. The building welcomes visitors to Georgia, representing the progressive development of the country.

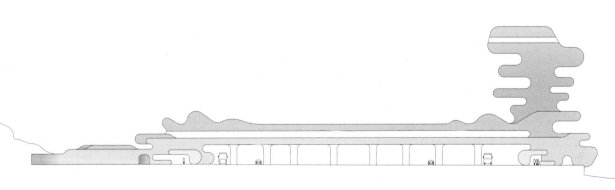

Sectional Drawing —— 剖面图

设计构思

悬臂式的平台，让塔楼具备了瞭望台的功能，许多层都可以俯视海面和海岸线陡峭的部分。 除配备常规的海关设施之外，建筑内还包括咖啡厅、员工休息室和会议室。建筑如一面喜迎八方游客的旗帜，象征着正在不断地发展的格鲁吉亚。

MARTIN LUTHER KIRCHE HAINBURG
马丁路德教堂
NEW COAT FOR CHURCH 时尚新衣

Location: Austria
Completion year: 2011
Designer: Coop Himmelb(l)au
Use: Church
Site Area: 420 m²
Client: Freunde der Evangelischen Kirche in Hainburg/Donau
Photographer: Duccio Malagamba

项目地址：奥地利
竣工时间：2011 年
项目设计：蓝天组建筑事务所
建筑用途：教堂
项目面积：420 m²
开 发 商：Freunde der Evangelischen Kirche in Hainburg/Donau
摄 影 师：Duccio Malagamba

Project Introduction

In less than a year a protestant church together with a sanctuary, a church hall and supplementary spaces was built in the centre of the Lower Austrian town Hainburg, at the site of a predecessor church that doesn't exist anymore since the 17th century. It's modelling avant-garde and unique changes the traditional church's solemn style.

项目介绍

不到一年的时间,这座崭新的清教徒教堂便出现在了奥地利的海恩堡。原址上的旧教堂,早在17世纪便废弃了。如今,新的教堂包括圣堂、教堂大厅以及其它辅助空间。它一改传统教堂肃穆的风格,造型前卫而独特。

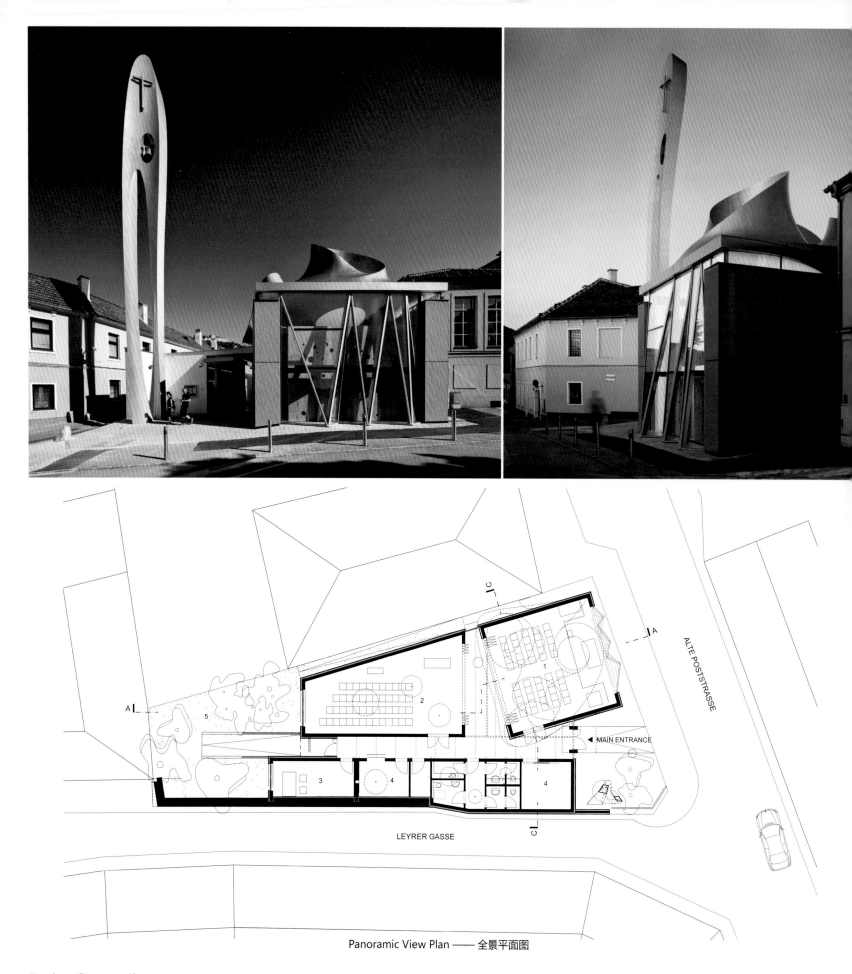

Panoramic View Plan —— 全景平面图

Design Conception

The shape of the building looks like a huge table. With its entire roof construction resting on the legs of the "table" – four steel columns. Another key element is the ceiling of the prayer room: its design language has been developed from the shape of the curved roof of a neighboring Romanesque ossuary – the geometry of this century-old building is translated into a form, in line with the times, via today's digital instruments. The play with light and transparency has a special place in this project. Facing the street, the glass curtauin wall is also attractive.

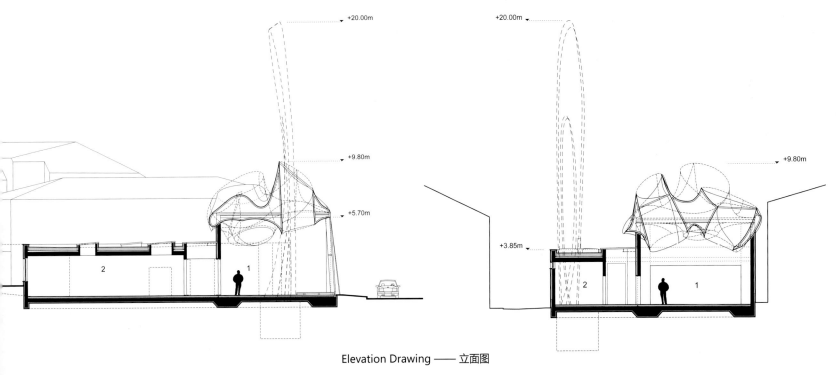

Elevation Drawing —— 立面图

设计构思

建筑的形状宛若一张巨大的桌子。整个屋顶都依靠"桌腿",即四个钢柱支撑。另一个重要的建筑元素是祈祷室上方的弧形屋顶,它的设计灵感源自邻近的罗马式藏骨堂。设计师用现代数字工具,为教堂塑造了一个符合时代的全新造型。它的主要特点是透明和开放。在流线型涡轮状的屋顶上,三个巨大的圆形天窗使得阳光一泻而下,照亮整个祈祷堂,宗教的神圣感在屋顶光源中得到渲染。而面向街道的曲折式玻璃幕墙,对游人也不乏吸引力。

IGUZZINI 照明公司总部
A HUGE BALL 巨大的球体

Location: Barcelona, Spain
Completion year: 2011
Designer: MiAS Architects
Use: Office Building
Site Area: 9,000 m²
Client: iGuzzini Illuminazione España SA
Photographer: Adrià Goula

项目地址：西班牙巴塞罗纳
竣工时间：2011 年
项目设计：MiAS Architects
建筑用途：办公
项目面积：9 000 m²
开 发 商：iGuzzini Illuminazione España SA
摄 影 师：Adrià Goula

Project Introduction

iGuzzini does not belong to the ground which it sits. Like a balloon, Leonidov's aerostat, it will attempt to escape from this world, seeking for a new sky. It will describe the conditions of the light, natural and artificial, in its interior, it will refer to its origins, recognizing a geometric order, but above all it tells us its aspirations.

项目介绍

iGuzzini 公司总部大楼仿佛不属于它脚下的这片土地，就像气球或者列昂尼多夫笔下的航空器一样，企图逃离这个世界，寻找一片新的天空。建筑内部充满自然光与人造光，并以几何序列进程介绍公司发展史，但最重要的是，它告诉了我们这个公司的抱负。

Design Conception

Our proposal seeks to exemplify, in architecture, the conditions closest to humankind including: collectivity, ambition, and excellence. iGuzzini is located at one of the roadway hubs of Barcelona, in a fragile condition of stability on the site where it sits. Without modifying the topography, a large underground space is previously delimited, providing storage, parking, a showroom, an auditorium, and an area for technical services. Its roof is proposed as an exterior technical floor such that it looks like an outdoor showroom. Atop this underground container emerges the company building, spherical in form though slightly deformed on the south side. A large central void occupied by the single column from which the entire building is suspended, permits greater light and energy control inside.

设计构思

设计旨在从建筑上体现人性化,比如:团结、雄心与卓越。建筑位于巴塞罗那的公路枢纽之一,场地不能经受较大改动。因此在保留原地形的条件下,建造了一个大型地下空间,包括储藏空间、停车场、陈列室、大礼堂以及设备用房。屋顶包含对外技术部,看起来就像室外展厅。球形建筑从地下空间拔地而起,南侧稍有变形。球体靠中心一系列结构支撑,远观仿佛悬浮在地下室上方,球形之外皆为无柱空间。因此有更多的自然光进入室内,有利于能源节约。

KUMUTOTO TOILET
KUMUTOTO 卫生间
RECALLING WATERFRONT'S SHIPPING PAST 忆往昔泊船岁月

Location: Wellington, New Zealand
Completion year: 2008
Designer: Studio Pacific
Use: Public Hygiene
Client: Wellington Waterfront Ltd.

项目地址：新西兰惠灵顿
竣工时间：2008年
项目设计：Studio Pacific
建筑用途：公共卫生
开 发 商：Wellington Waterfront Ltd.

Project Introduction

These public toilets are located at the Synergy Plaza in the Kumutoto precinct. As well as taking practical considerations into account such as security, hygiene and vandalism, the brief was to create a structure with a sculptural form, something iconic, highly visible and unusual that was also well integrated into the visual and historical context of the surrounding precinct.

项目介绍

该项目分布于 Kumutoto 区的协和广场。考虑到诸如安保、卫生及公物破坏等实际问题的同时，其设计旨在创建一个标志性的雕塑结构。独特的外观具有很高的辨识度，也很好地融入了周围的环境及文化氛围。

North Elevation —— 北立面

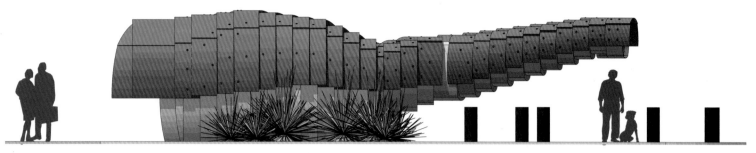

West Elevation —— 西立面

Design Conception

Throughout the panorama, the design comprises two elongated, irregularly curved parts, instantly recognisable from all key pedestrian approaches and terminating a sequence of spaces and elements along the laneway. The organic forms, eye-catching and instantly memorable, are suggestive of crustaceans or sea creatures, as if the structure was a kind of fossilised husk that had been discovered and inhabited.

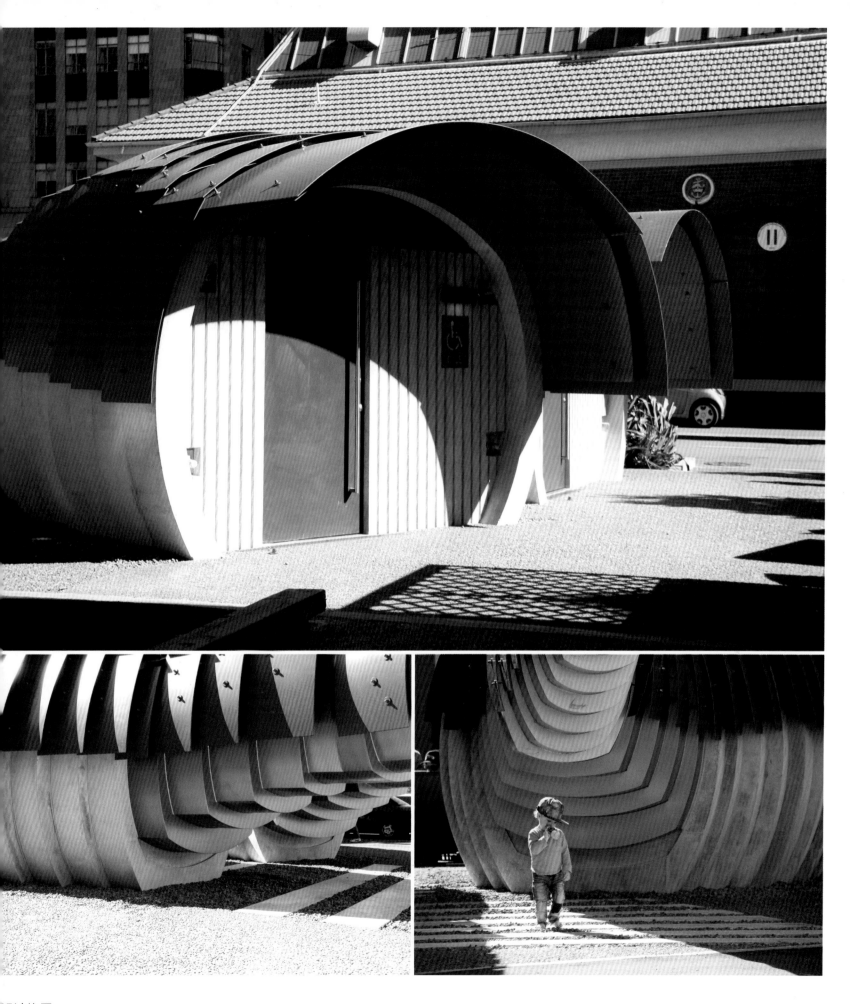

设计构思

纵观全景，该设计包括两个细长的、不规则的弧状结构。它们不仅易于识别，还在终端留出一系列空间，为路面提供了装饰元素。奇特的造型让人不禁联想起某种甲壳类生物或是海洋生物，整个建筑就像是出土后的化石被发现和再利用了一般。

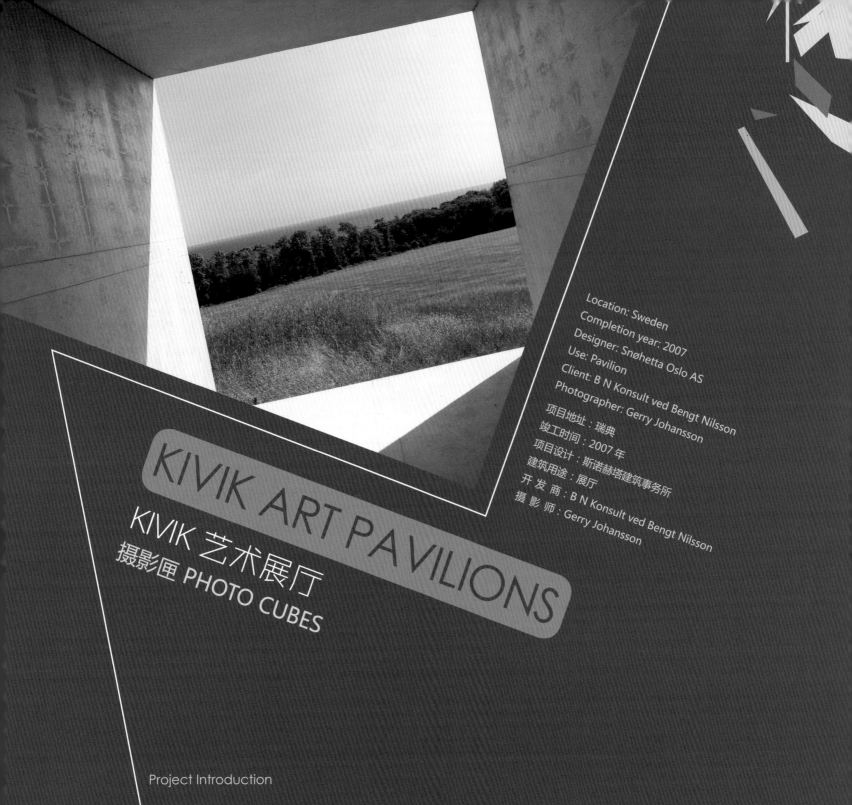

KIVIK ART PAVILIONS
KIVIK 艺术展厅
摄影匣 PHOTO CUBES

Location: Sweden
Completion year: 2007
Designer: Snøhetta Oslo AS
Use: Pavilion
Client: B N Konsult ved Bengt Nilsson
Photographer: Gerry Johansson

项目地址：瑞典
竣工时间：2007 年
项目设计：斯诺赫塔建筑事务所
建筑用途：展厅
开 发 商：B N Konsult ved Bengt Nilsson
摄 影 师：Gerry Johansson

Project Introduction

It is the first phase in the establishment of a new contemporary art centre in Kivik, Sweden. Snøhetta was invited to design a temporary pavilion that combines art and architecture together with an artist of their unique feature. It is a series of explorations of the spatial and temporal relationship between photography, architecture and the landscape.

项目介绍

该项目是新当代艺术中心的一期工程。设计师 Snohetta 应邀设计这个临时展厅。设计要求在艺术和建筑融合的同时,体现参展摄影师的特色。这无疑是处理摄影、建筑和景观三者关系的探索性课题。

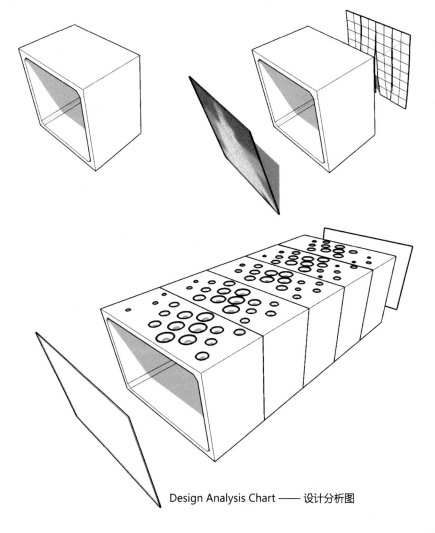

Design Analysis Chart —— 设计分析图

Design Conception

Strategic placement of the 5 concrete interventions on the site invites the visitor to explore four discrete landscape experiences. Two viewfinders are 2.5x2.5m cubes that are open on two sides. The Mothership is made of a series of five 3x5m rectangular concrete elements which functions as a flexible exhibition space. Two Photo-boxes are 2.5 x 2.5m cubes with a Tom Sandberg photograph silk printed on laminated glass on the other side with a black neoprene cover on the back stare at each other across a bend in the road.

设计构思

该项目包括 5 个分散的混凝土建筑，它们为游客带来了四种不同的观景体验。被命名为"取景器"的是两个 2.5x2.5 m 的立方体建筑，两端是开放的。被命名为"母舰"的是由 5 个 3x5 m 矩形混凝土元件组成的、灵活的展览空间。被命名为"相片匣"的是两个 2.5 x 2.5 m 立方体建筑，一面是印有 Tom Sandberg 摄影作品的叠层玻璃，另一面是黑色的氯丁橡胶。两个"相片匣"隔街相望。

TREEHOUSE DJUREN
DJUREN 树屋
GREEN TRIP 绿野仙踪

Location: Germany
Completion year: 2008
Designer: Baumraum
Use: Rest & Play
Site Area: 16.4 m²
Photographer: Alasdair Jardine

项目地址：德国
竣工时间：2008 年
项目设计：Baumraum
建筑用途：休息、娱乐
项目面积：16.4 m²
摄 影 师：Alasdair Jardine

Project Introduction

Treehouse Djuren was designed by Baumraum. It's extraordinary and comfortable, and it serves as a room for rest and relaxation for the adults as well as a room to play for the children. The treehouse-construction is devided into two parts: the lower terrace and the treehouse with a small terrace.

项目介绍

该项目舒适而特别,是大人休闲、孩童玩耍的绝佳之地。它从结构上分为两个部分——低处的平台和带阳台的树屋。

Rendering —— 效果图

Design Conception

The rounded shape of this treehouse is reminiscent of an egg which was cut open longitudinally. This association is heightened through the accenting of the gable surfaces with cream-painted perspex, and the elliptically-shaped windows. On the other hand, the materials chosen for the other external elements, such as the terrace and the underside of the treehouse, are more robust, with these being constructed of indigenous oak. Sheet zinc was used for the treehouse roof. One special detail is the curved glass area on the front façade. The weight of the treehouse is borne by both the trees and support frame. The weight of the two terraces and the horizontal load of the treehouse are distributed across the oaks by means of steel cables and textile straps.

Elevation Drawing —— 立面图

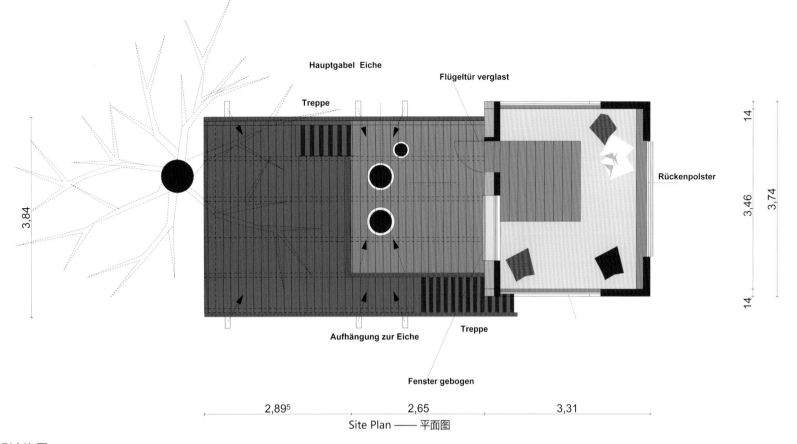

Site Plan —— 平面图

设计构思

树屋的圆润外形让人不禁联想到被纵向切开的鸡蛋。奶油色透明塑胶的三角墙和椭圆形的玻璃，更是突出了"鸡蛋"的特征。外立面其他部分的选材，例如平台和树屋下部，都选用了原生态的本土橡木。屋顶由锌薄皮制成。表皮前部的玻璃制曲面是特别的细节之一。树屋的重量由树干和自身的支撑架一同承载。不仅如此，为了减少房子对树的压力，两个平台的重量和树屋的水平负载分散在若干根支柱上，靠坚韧的钢绳和纺织绳带来平衡，保证绝对的稳定性，且不对树木造成伤害。

TEA HOUSE
茶亭
OTHERWORLDLY MEDITATION 冥想之地

Location: Maryland, USA
Completion year: 2009
Designer: David Jameson Architect, Inc.
Use: Muti-purposed
Photographer: Paul Warchol

项目地址：美国马里兰州
竣工时间：2009 年
项目设计：David Jameson Architect, Inc.
建筑用途：多功能
摄影师：Paul Warchol

Project Introduction

A hanging bronze and glass object inhabits the backyard of a suburban home. The structure, which evokes the image of a Japanese lantern, acts as a tea house, meditation space and stage for the family's musical recitals.

项目介绍

该项目位于市郊一座房子的后院,它由青铜和玻璃制成,凌空悬挂于草坪上。它的造型独特,引起人们对日本传统灯笼的遐想,另外它的功能丰富,除了能让人在这里享受品茶的乐趣,还能成为一个好的冥想场所和家庭聚会时表演的舞台。

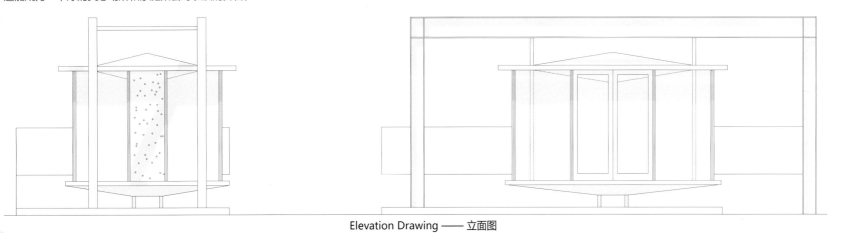

Elevation Drawing —— 立面图

Design Conception

After experiencing the image of the lantern as a singular gem floating in the landscape, one is funneled into a curated procession space between strands of bamboo that is conceived to cleanse the mind and prepare one to enter the object. After ascending an origami stair, the visitor is confronted with the last natural element: a four inch thick, opaque wood door. At this point the visitor occupies the structure as a performer with a sense of other worldliness meditation.

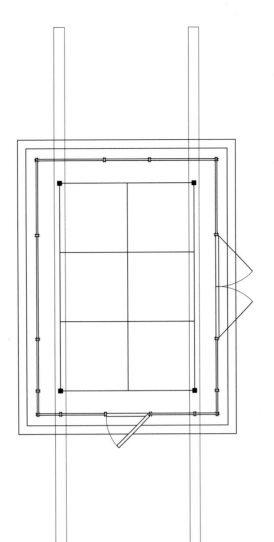

Site Plan —— 平面图

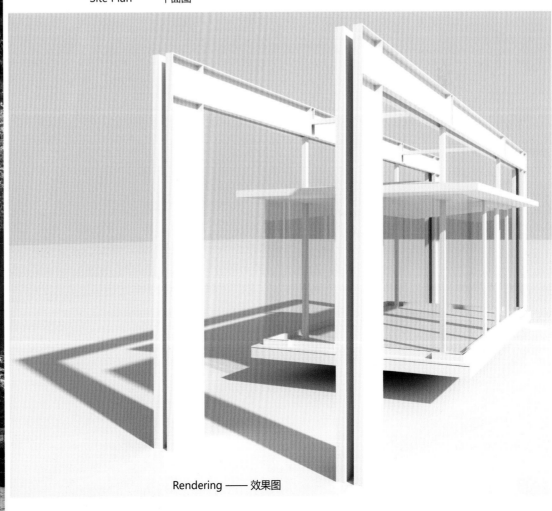

Rendering —— 效果图

设计构思

优美典雅的外形使它犹如一颗漂浮的宝石。当人们迈入竹林,见到这独特的美景时,仿佛心灵瞬间得到了净化,同时也萌生想要进入一探究竟的念头。拾级而上,通过那道四英尺厚的实木门后,人们会有一种神奇的体验,就像艺术家进入了最理想的冥想空间。

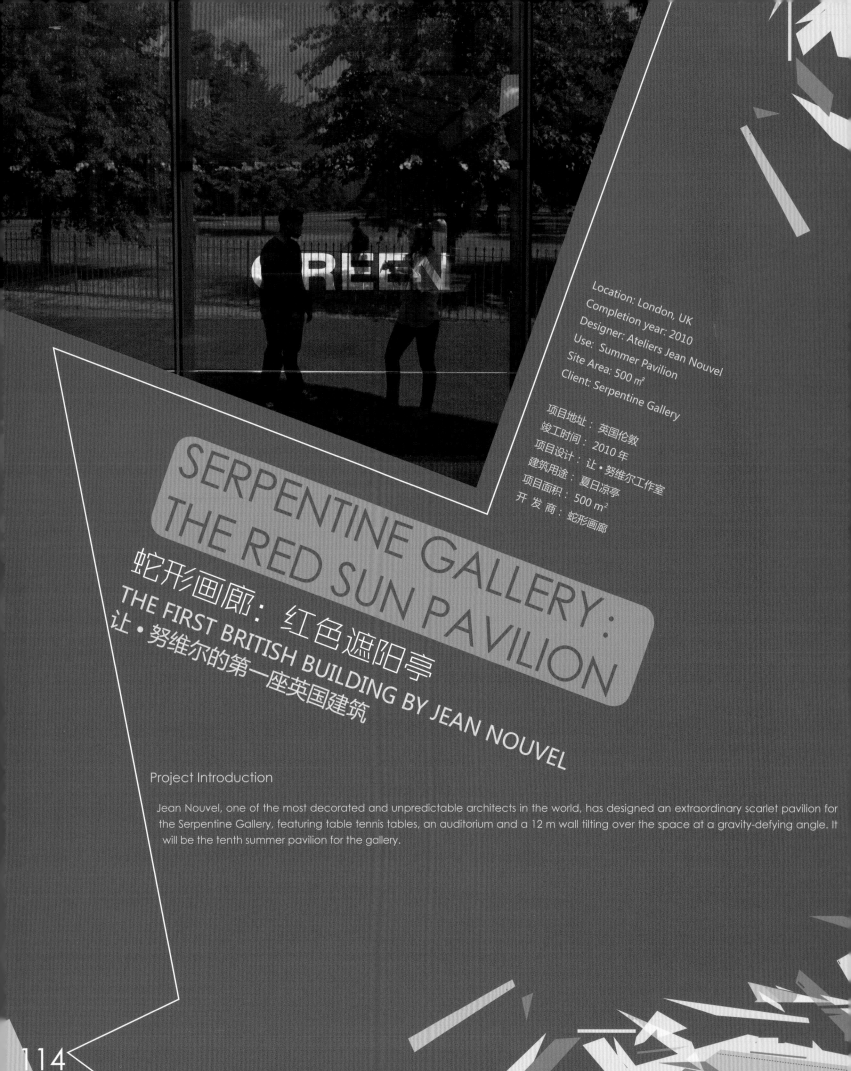

SERPENTINE GALLERY: THE RED SUN PAVILION
蛇形画廊：红色遮阳亭
THE FIRST BRITISH BUILDING BY JEAN NOUVEL
让·努维尔的第一座英国建筑

Location: London, UK
Completion year: 2010
Designer: Ateliers Jean Nouvel
Use: Summer Pavilion
Site Area: 500 m²
Client: Serpentine Gallery

项目地址：英国伦敦
竣工时间：2010 年
项目设计：让·努维尔工作室
建筑用途：夏日凉亭
项目面积：500 m²
开 发 商：蛇形画廊

Project Introduction

Jean Nouvel, one of the most decorated and unpredictable architects in the world, has designed an extraordinary scarlet pavilion for the Serpentine Gallery, featuring table tennis tables, an auditorium and a 12 m wall tilting over the space at a gravity-defying angle. It will be the tenth summer pavilion for the gallery.

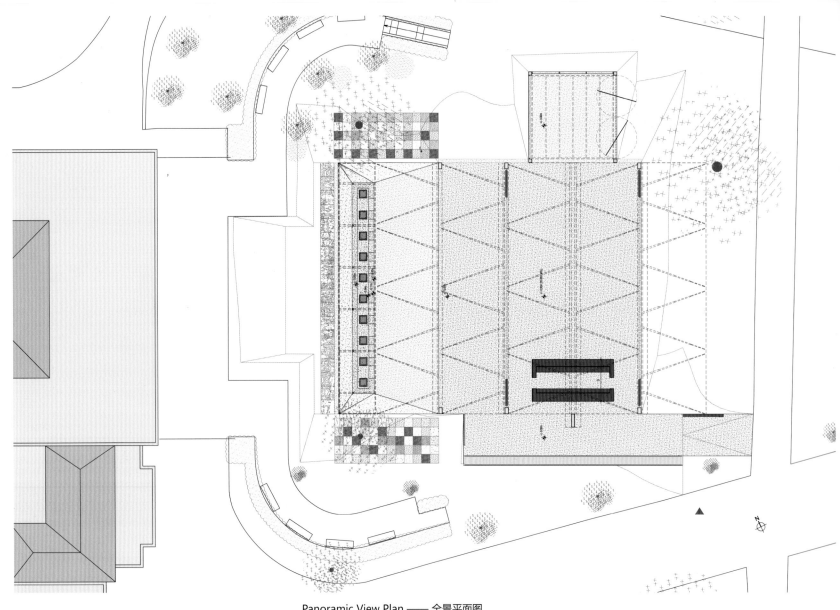

Panoramic View Plan —— 全景平面图

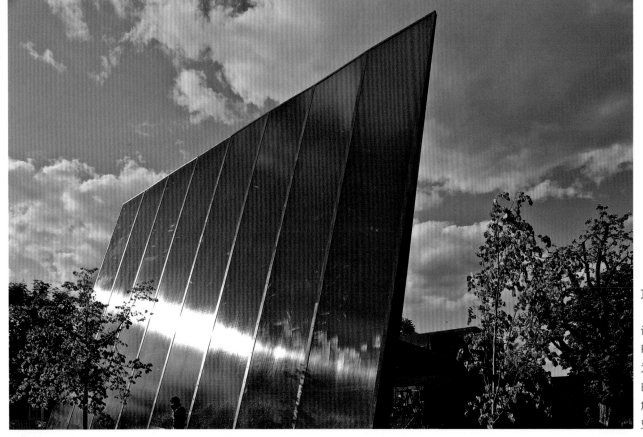

项目介绍

让·努维尔，这位普利兹克奖 (Priztker Prize) 获得者、风格独特的法国建筑大师为蛇形画廊设计了一座鲜红色的建筑。它由户外乒乓球桌、礼堂和反重力形成倾斜角的 12 m 墙体构成。这将是蛇形画廊有史以来建立的第十个夏日凉亭。

Design Conception

Jean Nouvel's pavilion is constructed up of glass, polycarbonate and fabric structures. Made up of cantilevered parts and retractable awnings, the bold geometric forms of the building are rendered in a bright red, contrasting the natural park setting in which it is situated. The color reflects iconic British images such as telephone boxes, postbox and London buses. The exterior and interior spaces are designed in a flexible manner to accompany different numbers of guests and the changing of the weather.

设计构思

该设计是轻型材料和金属悬臂式结构的结合，以画廊的40周年纪念为主题，呈现伦敦红，与郁郁葱葱的景观形成了鲜明的对比。建筑的颜色与英国的经典电话亭、邮筒、城市巴士这些标志性形象颜色一致。但更多原因，设计师选择红色是因为太阳。他想做的当然不仅仅是反映太阳的光辉，更希望它能够捕捉和过滤情绪，成为一个温暖快乐的小空间。最令人惊奇的是画廊的外形可以根据游客的数量和天气的变化而调整，创造不同的内外空间。

PARTY ANIMAL
派对动物
RED STAGE 红色舞台

Location: Lisbon, Portugal
Completion year: 2011
Designer: LIKEarchitects
Use: Stage Site
Area: 80 ㎡
Client: Ordem dos Arquitectos
Photographer: Francisco Nogueira

项目地址：葡萄牙里斯本
竣工时间：2011 年
项目设计：LIKEarchitects
建筑用途：舞台
项目面积：80 m²
开 发 商：Ordem dos Arquitectos
摄 影 师：Francisco Nogueira

Project Introduction

The 'Party Animal' intervention took up stage in an attractive, yet underused, passageway, with the aim of maximizing the celebrations of the patron Saint António. The festival of Santo António is citywide, month long, carnival.

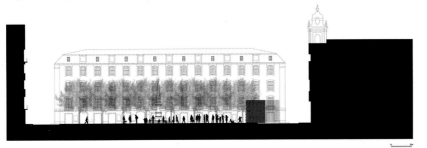

Elevation Drawing —— 立面图

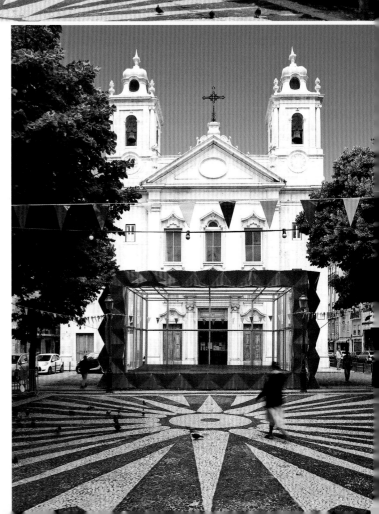

项目介绍

该设计将舞台搬到了极具吸引力、却未被充分利用的人行巷道,旨在为 Saint António 欢庆活动添彩助兴。这个全市范围的节日为期一个月,以嘉年华活动的方式展开。

Design Conception

The idea was that the intervention would focus mainly on an attractive stage to be the platform for the invited artists to act. It should be a single low-cost object with the ability to be the engine of this new centre of cultural events - a temporary piece that would be the star both of the square and the parties, made to receive not only popular concerts, but also rock, world music, fado nights and DJs. Despite the scarce resources and the huge volumetric scale desired for the stage, this lightweight structure (made out of prefabricated modular parts) was designed to impose itself as an iconic urban catalyser while retaining and respecting the consolidated spatial hierarchy of the square. At the same time, it is a project that takes advantage of the use of colour as an urban activator contrasting with the subdued colour of the old town.

设计构思

这个极具魅力的舞台，是艺术家表演的绝佳场所。虽然它的制作成本不高，但是却能引领整个文化潮流。它不仅是广场的焦点，也是聚会的中心。在这里将上演流行音乐会、摇滚派对、演奏民族音乐、葡萄牙民谣 fado 等。尽管绚丽的舞台离不开稀有材料和大型体量，这个用预制模块搭成的轻质结构还是以其独有的外观成为了城市的焦点。它保留广场空间固有的层次，同时，利用鲜艳的颜色，调和了周围的环境，与老城区的背景色彩形成了鲜明的对比。

RINGS AROUND A TREE
树之环
A CLASSROOM WITHOUT FURNITURE 没有桌椅的教室

Location: Tokyo, Japan
Completion year: 2011
Designer: Tezuka Architects
Site Area: 46.98 m²
Client: Montessori School Fuji Kindergarten
Photographer: Katsuhisa Kida

项目地址：日本东京
竣工时间：2011年
项目设计：Tezuka Architects
项目面积：46.98 m²
开 发 商：蒙特梭利学校富士幼儿园
摄 影 师：Katsuhisa Kida

Project Introduction

The annex to Fuji Kindergarten contains English classrooms and a school bus station. A twisting zelkova tree dominates the site and while half of the building is exterior space, the footprint does not define the boundary between outside and in.

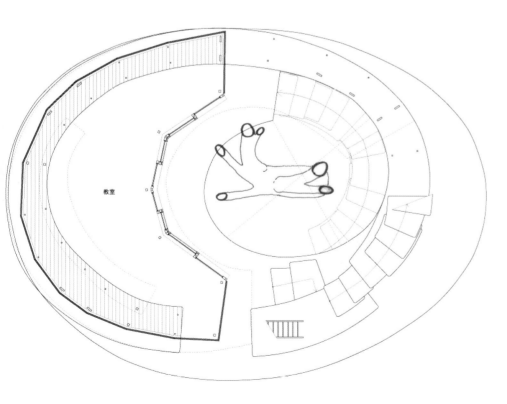

Site Plan —— 平面图

项目介绍

该项目是富士幼儿园的附加设施,不仅可以作为英语教室使用,同时也是该幼儿园的校车车站。建筑物中心被蔓生的榉树枝杈和茂密的树叶穿透,打破了室内与室外、建筑与自然环境之间的界限。

Design Conception

The oval-shaped plan traces the zelkova's broad canopy making the columns and floor seem to vanish in the shimmering shadows. Existing branches take precedence and penetrate the building, and grown-ups have to crawl when ascending the stairs to the roof. The angled trunk is perfect for climbing, with bark polished smooth by generations of small, adventuresome hands. Previously, a tree house occupied the site, so small that only children could enter. What the designer could not ignore was when children began climbing over the handrail and out onto the branches of the tree, this was dangerous. They solved this issue by tying ropes around certain areas.

设计构思

交错排列的楼板小心地包围着中心大树,建筑内部的柱子和楼梯隐藏在树叶和枝桠的阴影中。上楼的过程变得非常有趣!这个巨大的开放式"树屋"中有许多压缩和封闭的空间,只有半蹲和爬行的小朋友才能通过。为了保护在建筑中活动的小朋友,设计师没有设计过多的障碍物和栏杆,而是在地面上覆盖了一层橡胶垫,并在主要的攀爬区域围上绳子,这样可以减轻一些意外磕碰对孩子们造成的伤害。

BURNHAM PAVILION
伯纳姆展亭
DISPLAY THE OUTLINE 展示城市轮廓

Location: Chicago, USA
Completion year: 2010
Designer: UNStudio Architecture
Use: Pavilion

项目地址：美国芝加哥
竣工时间：2010 年
项目设计：UNStudio 建筑事务所
建筑用途：展览

Project Introduction

Placed on a unique location in the middle of Millennium Park and framed by Lake Michigan on one side and Michigan Avenue on the other, the UNStudio pavilion relates to diverse city-contexts, programs and scales. Programmatically the pavilion invites people to gather, walking around, exploring and observing.

Views and Orientation

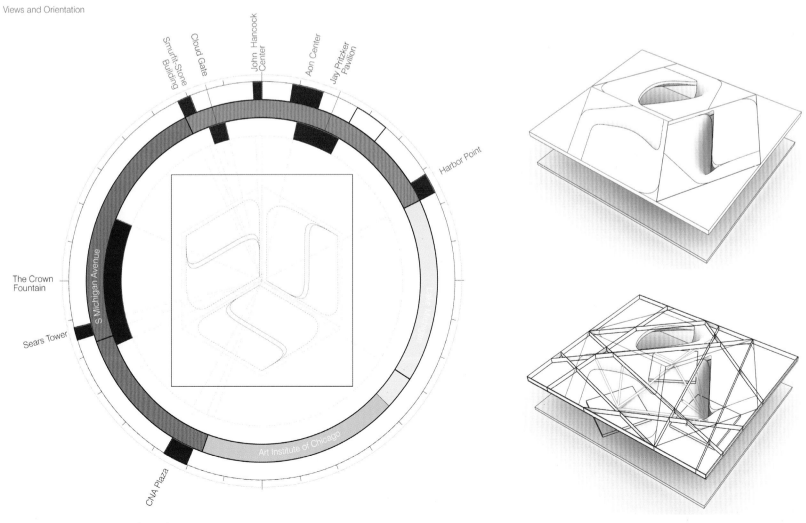

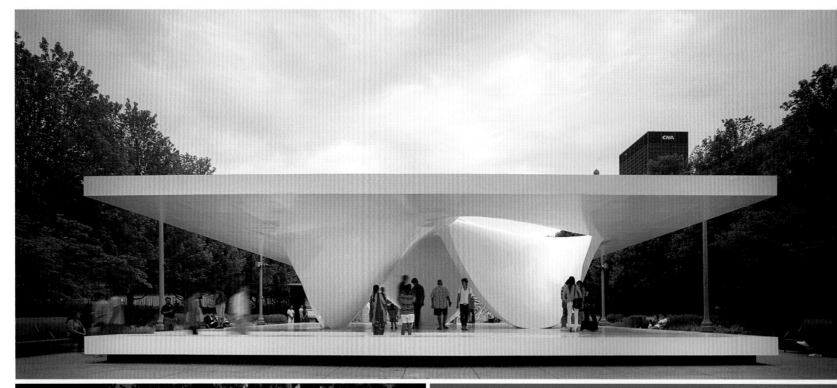

项目介绍

该项目地理位置独特，位于千禧公园的中心，一面是密歇根湖，另一面是密歇根大道。它联系着不同的城市环境、规划项目和建筑空间。独特的造型吸引人们聚集于此，或漫步观赏，或进去一探究竟。

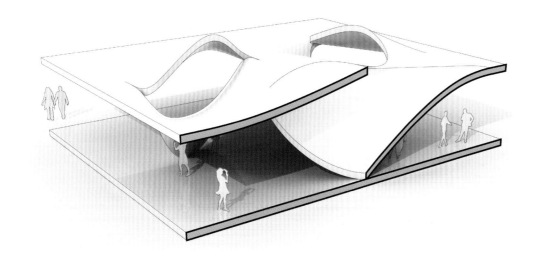

Design Conception

The UNStudio pavilion is sculptural, highly accessible and functions as an urban activator. Based on the specificity of the site, the design of the pavilion elaborates on the relationship to the existing rigid geometry, but it also introduces a floating and multi-directional space. The pavilion is open at its both two sides, between the two horizontal planes of the podium and roof. It orients itself to the city texture, to the flows of visitors exploring Millennium Park and most importantly introduces diverse vistas towards the park and city surroundings.

设计构思

该项目是一个装饰性建筑，外形亲切，为城市注入了活力。以特殊的地理位置为基础，设计阐明了现存的几何关系，且引入了一个流动的、多方向的空间。墩座墙和屋顶的两个水平面之间是开放的。该建筑服务于城市，服务于参观公园的人群，最重要的是，它为公园和城市周围景观带来了不同的视角。

ON THE CORNER
街角住宅
BLUE CROSS 蓝色交叉

Location: Shiga, Japan
Completion year: 2011
Designer: EASTERN Design Office
Use: Residence
Site Area: 261 m²
Client: TOYO-KAIHATSU Co., ltd
Photographer: Koichi Torimura

项目地址：日本滋贺
竣工时间：2011 年
项目设计：EASTERN Design Office
建筑用途：住宅
项目面积：261 m²
开 发 商：TOYO-KAIHATSU Co., ltd
摄 影 师：Koichi Torimura

Project Introduction

The site is a manufacturing area in Youkaichi City in Shiga Prefecture. There are many big factories in this town. A lot of immigrants from South America live here among the local people.This is a residential area as well an industrial area. It is also a popular drinking area where many bars and restaurants for such common people are scattered. There is a highway interchange nearby the town. The shape of the lot in a triangle deeide there is in unusual building.

Site Plan —— 平面图

项目介绍

该项目位于日本八百日市滋贺县的工业住宅区，居住着许多来自南美的移民，聚集着众多酒屋和餐馆。由于它建在道路交汇处一个尖锐的三角区域，所以决定了房子会拥有不同寻常的外形。

Design Conception

It looks like a present, a toy box or a castle where the boys and girls in the story of Michael Ende could be entering. It is a triangular building configured by the square elements. The cross confines the power of the mixed materials into one. A shuffle of stones, concretes, and glass. Keen edge of each material is too sharp. The design "on the corner" consists of blue and the cross. It seems as if this illusion deceives people to obscure their eyesight and feel invited to another world. It is pretentious, yet it is surrealistic too.

设计构思

该项目看上去就像一件礼物，一个玩具盒子或是童话作家 Michael Ende 书中的城堡。设计师将自己的设计概括为：混凝土、石材和玻璃的混合，清晰与锐利表现在每一条边缘与每一种材质上。设计师试图利用线与面的组合，突出剥离于城市框架的形状。建筑的结构给人一种错觉，如果它能欺骗人的视觉，让人感觉身处另一个世界，那么它便是超凡脱俗且富有超现实主义色彩的建筑。

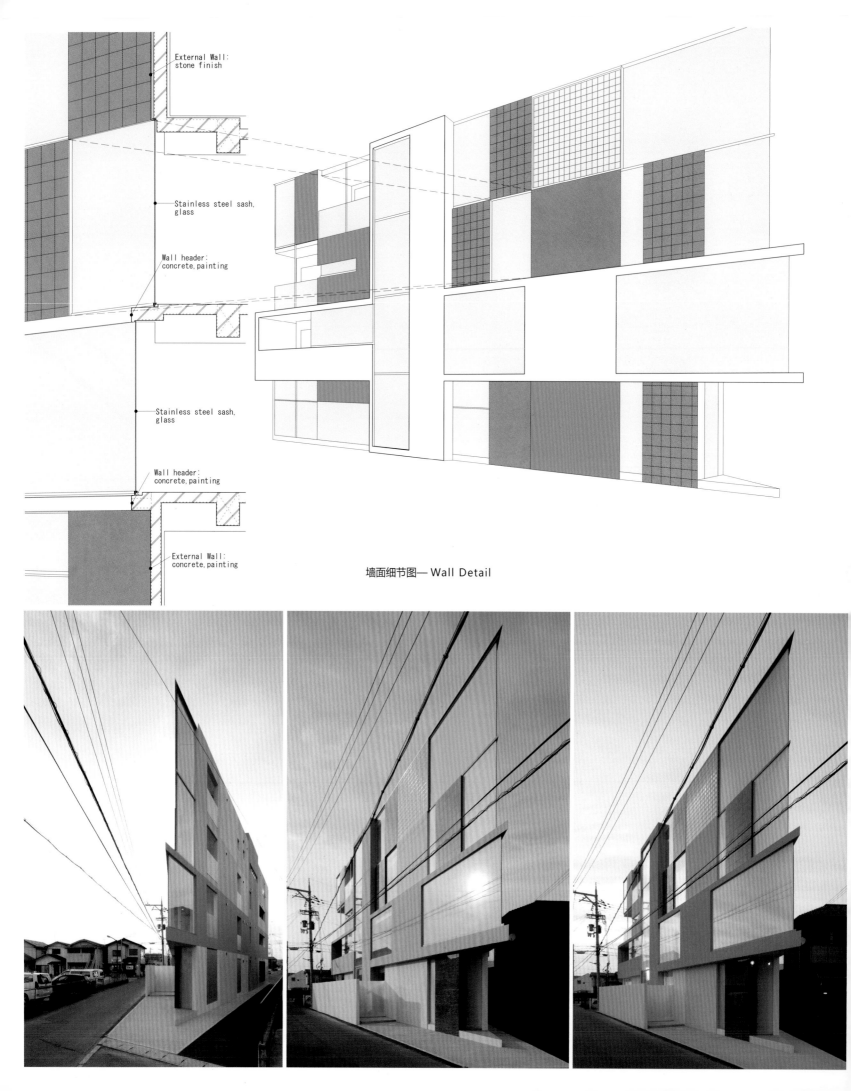

墙面细节图— Wall Detail

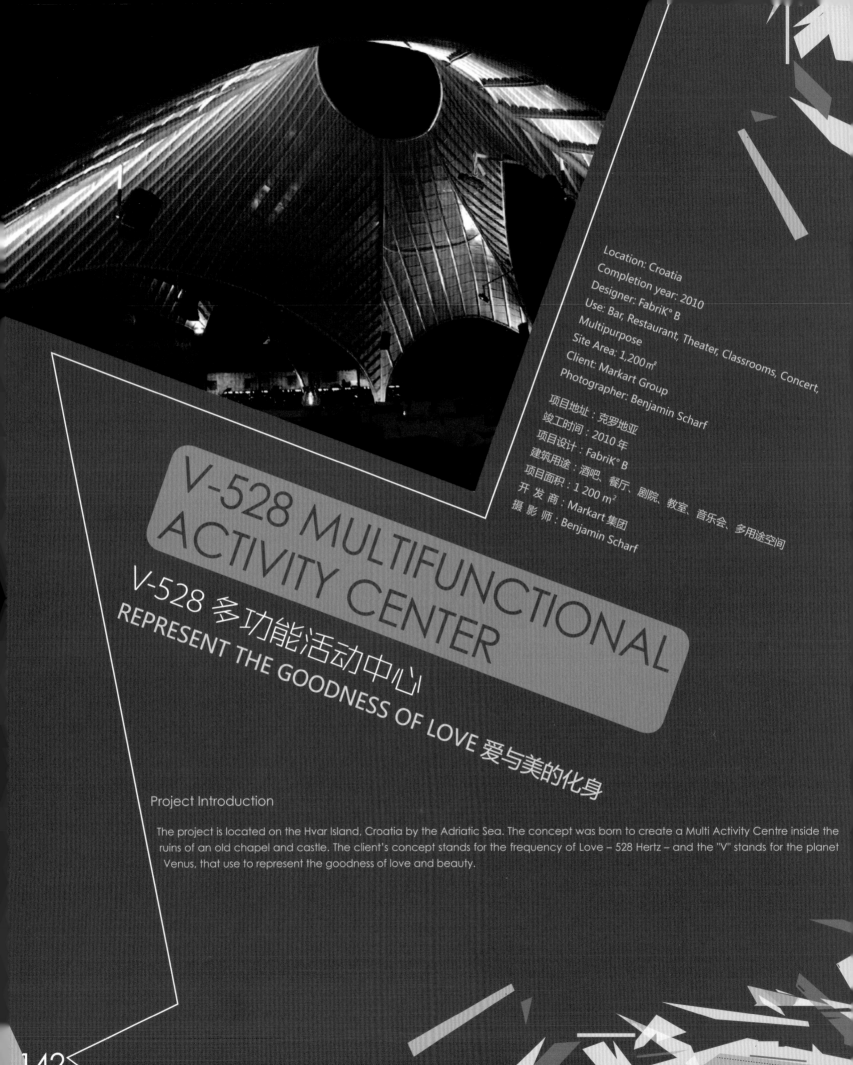

V-528 MULTIFUNCTIONAL ACTIVITY CENTER
V-528 多功能活动中心
REPRESENT THE GOODNESS OF LOVE 爱与美的化身

Location: Croatia
Completion year: 2010
Designer: FabriK° B
Use: Bar, Restaurant, Theater, Classrooms, Concert, Multipurpose
Site Area: 1,200㎡
Client: Markart Group
Photographer: Benjamin Scharf

项目地址：克罗地亚
竣工时间：2010 年
项目设计：FabriK° B
建筑用途：酒吧、餐厅、剧院、教室、音乐会、多用途空间
项目面积：1 200 m²
开 发 商：Markart 集团
摄 影 师：Benjamin Scharf

Project Introduction

The project is located on the Hvar Island, Croatia by the Adriatic Sea. The concept was born to create a Multi Activity Centre inside the ruins of an old chapel and castle. The client's concept stands for the frequency of Love – 528 Hertz – and the "V" stands for the planet Venus, that use to represent the goodness of love and beauty.

Panoramic View Plan —— 全景平面图

项目介绍

该项目位于赫瓦尔岛，亚得里亚海附近。这是一个在废弃的老教堂和城堡内建造的多功能活动中心。它名字内的数字来源于惯用的代表爱的频率 528 Hz，而 "V" 代表金星维纳斯——爱与美的化身。

Design Conception

The designer took the basic idea and implement new roof structure that organizes the program together with flexibility but creating different roofed areas. At the same time all the open space is visually connected in a new way and in the other hand brings independence also with different platforms without losing the visual communication.The forms are also created out of the different needs of the concept, like the Theatre Dome that is shaped by visual and functional patterns. Or the dance floor which is also was yoga, "capoeira" and other classes. Unify the 3 domes in one. The structure is made wp of two layers of steel-tubes that are connected with each other as a spider web does.

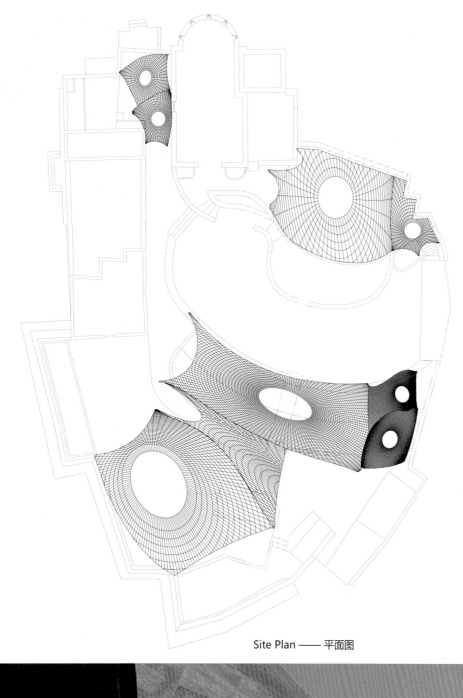

Site Plan —— 平面图

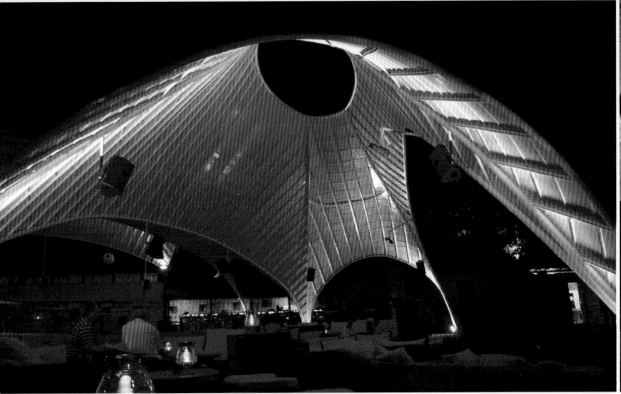

设计构思

设计师贯彻了基本的理念要求，制造了一个新的屋顶结构，将各种空间灵活地统一起来。不同的舞台具有各自的独立性，但相互之间不失视觉联系。这样的形式也创造出多样的功能。如圆顶电影院在视觉和功能上都很出众；在舞蹈室既可以做瑜伽，也可以跳卡泼卫勒舞，或是开展其它课程活动。三个穹顶连为一体。而两层钢管构成的结构，交错连接，就像是一张巨大的蜘蛛网。

TIGER AND TURTLE - MAGIC MOUNTAIN
老虎与乌龟：魔山
WALK ON ROLLER COASTER'S TRACK 徒步式过山车

Location: Duisburg, Germany
Completion year: 2008
Designer: Ulrich Genth, Heike Mutter
Use: Large-scale Sculpture

项目地址：德国杜伊斯堡
竣工时间：2008 年
项目设计：Ulrich Genth 和 Heike Mutter
建筑用途：大型雕塑

Project Introduction

The sleek curved shape of a rollercoaster highlights widely visible the highest peak of the park. The visitor can climb the art work by foot. Although the course describes a closed loop, it is impossible to accomplish it as the looping emerges to be a physical barrier.

该项目以其独特的过山车造型,屹立在公园山峰的制高点,成为了整个区域的焦点。与普通的雕塑不同,这个作品可以供游人攀爬。尽管整个轨道都设有扶栏,但由于超越了物理极限,人类是无法走完全程的。

项目介绍

该项目以其独特的过山车造型,屹立在公园山峰的制高点,成为了整个区域的焦点。与普通的雕塑不同,这个作品可以供游人攀爬。尽管整个轨道都设有扶栏,但由于超越了物理极限,人类是无法走完全程的。

Design Conception

"Tiger and Turtle" refers with its immanent dialectic of speed and deadlock to the situation of change in the region and its turn towards renaturation and restructuring. The dynamic sweeps and curves of the construction inscribe themselves like a signature into the scenery. From a distance the metallic glossy track creates the impression of speed and exceeding acceleration. Viewed nearly up, the supposed lane turns out to be a stairway which, elaborately winding, follows the course of the rollercoaster.

设计构思

项目的命名源自轨道不同区域所能提供的速度——随着轨道的不同高度和坡度，过山车时而如老虎般奔驰，时而如乌龟般爬行。雕塑的动态美为周围的美景画龙点睛。从远处看去，光滑的金属轨道给人过山车般呼啸而过的错觉，极具震撼力。走到近处，才发现其上有供攀爬的台阶，弯曲盘旋。

FOOTBRIDGE IN EVRY
埃夫里人行桥
FUNCTION FIRST 实用为本

Location: Evry, France
Completion year: 2007
Designer: DVVD
Use: Footbridge
Client: AFTRP
Photographer: DVVD / Alain Philippe Baudry Knops

项目地址：法国埃夫里
竣工时间：2007 年
项目设计：DVVD
建筑用途：人行桥
开 发 商：AFTRP
摄 影 师：DVVD / Alain Philippe Baudry Knops

Project Introduction

Due to the separation of pedestrian and vehicular traffic, bridges are particularly significant things in public space in Evry.

项目介绍

桥梁在埃夫里市既有实用价值——划分行人通道与车辆通道,也具有极高的装饰价值。

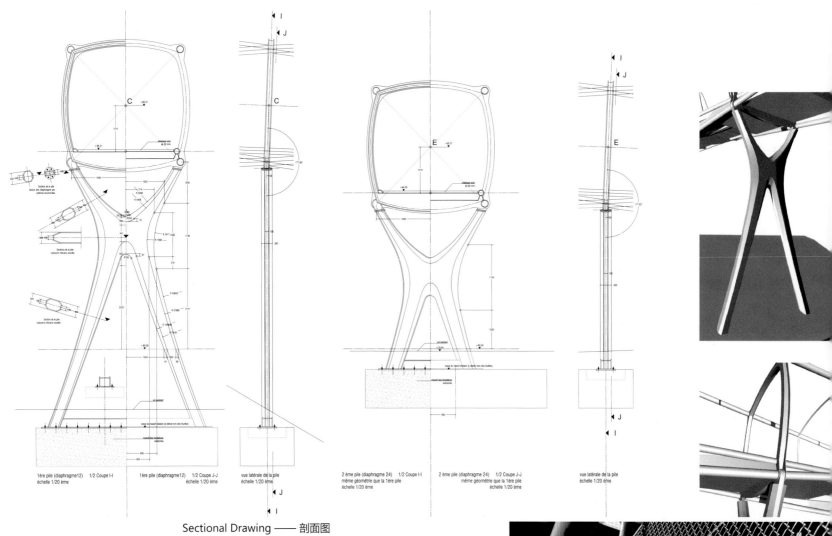

Sectional Drawing —— 剖面图

Design Conception

One key requirement quickly emerged: the need to prevent objects being thrown doun vehicles beneath the footbridge. The designer combined the architectural intentions and technical considerations to create a steel structure forming a volume which encompasses pedestrians. This loadbearing structure consists of an array of round tubes echoing the rotation of DNA across the entire span of the footbridge.

设计构思

设计团队首先需要通过设计,解决行人向桥下乱扔杂物的问题。结合建筑意向和工艺考虑,他们最终创造了一个可以环绕行人的钢铁结构。这个具有承载力的结构由大量的环状管沿着桥缠绕而成。

WOVEN BRIDGE
编织的桥
MODERN DESIGN 设计新颖

Location: Copenhagen, Denmark
Completion year: 2012
Designer: MLRP Architecture
Use: Bridge

项目地址：丹麦哥本哈根
竣工时间：2012 年
项目设计：MLRP 建筑事务所
建筑用途：桥梁

Project Introduction

The Woven Bridge is a modern interpretation of a classical steel park bridge. It allows for viewing the park and lake from new perspectives. The bridge is placed at the southern end of the lake and makes a new shortcut, when crossing through the park.

项目介绍

该项目用现代手法重新诠释了传统的公园钢制桥。它为公园和湖泊提供了新的观景点。桥梁坐落于湖泊的南端,是穿行公园的新捷径。

Design Conception

Designing the bridge, the key was to create a bridge that would blend into the landscape of the park but at the same time create a structure that has its own identity and personality. The bridge joins the two banks of the lake, which are at different heights, with a slender arched structure. The foundations are concealed under the steel structure making the transition between bridge and nature more refined. The bridge gets its name from its steel railing, which resembles a continuous woven thread, and gives the bridge a refined level of detailing, which is inspired from the classical steel railings and gives the structure a simplified ornamentation appropriate to its surroundings.

设计构思

将结构独特的桥梁完美融入公园景观中是设计的关键。桥梁连接着高度不一的湖泊两岸,形成细长的弓形结构。隐藏于钢铁结构下的底座,使桥体自然过渡。桥梁的命名来源于它类似编织线的栏杆。它们不仅是受传统铁栏启发后改良的精致细部,还是易于融入周围环境的简约装饰。

ZAPALLAR PEDESTRIAN BRIDGE
札帕拉尔人行天桥
A BEAUTY QUEEN TIARA 女王之冕

Location: Zapallar, Chile
Completion year: 2009
Designer: Enrique Browne
Type: Bridge
Client: Zapallar Municipality
Photographer: Tomás Rodriguez

项目地址：智利札帕拉尔
竣工时间：2009 年
项目设计：Enrique Browne
景观类型：桥梁
开 发 商：札帕拉尔市政府
摄 影 师：Tomás Rodriguez

Project Introduction

The bridge was projected in laminated wood and it is simply supported in concrete bases. Its execution and assembly was made partly in Santiago and moved to Zapallar, where it was installed and finished.

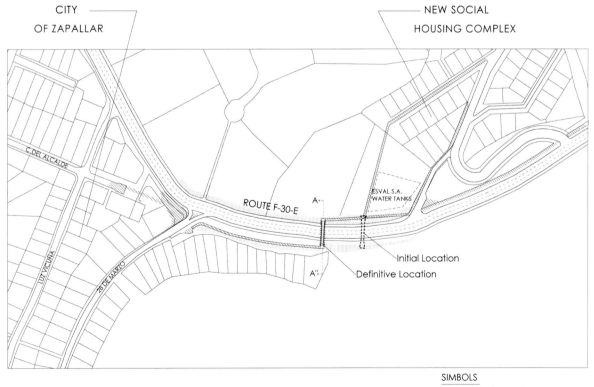

SITUATION PLAN

SIMBOLS
- Pedestrian walkways
- Green slope
- Existent retaining wall

Panoramic View Plan —— 全景平面图

项目介绍

桥身采用了被混凝土底座支撑、多层胶合板向外伸展的造型。部分的施工和组装是在圣地亚哥执行的，然后运至札帕拉尔，进行最后的安装和收尾工作。

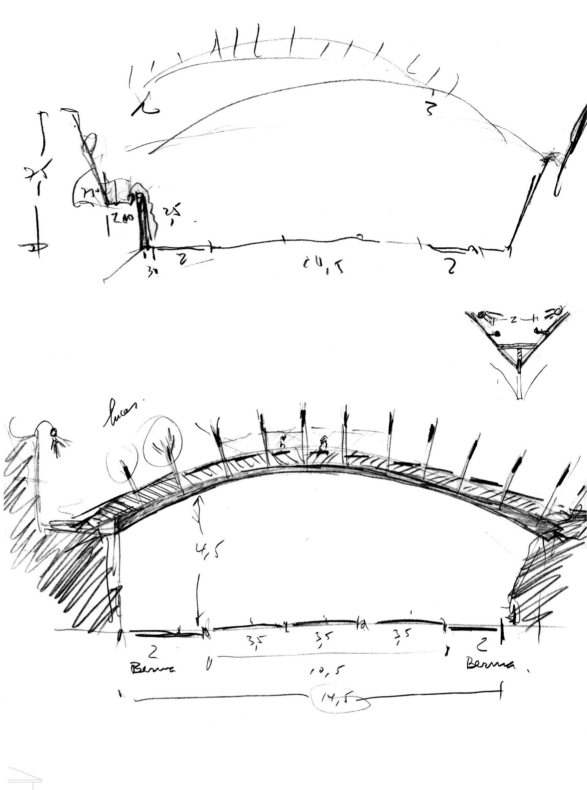

SOUTHERN ELEVATION

South Elevation——南立面

Design Conception

The designers were commissioned to make a bridge that would bring security to pedestrians, connecting for the high part both sectors. The bridge would also serve to transfer water pipes, electricity, etc.

设计构思

这座桥梁连接着两座小山,其设计旨在为行人带来安全而便利的通道。不仅如此,它还可以传输水电等。

RIGNY BRIDGE
RIGNY 桥
WHITE BEAUTY 白色丽影

Location: France
Completion year: 2011
Designer: DVVD
Use: Bridge
Client: Town of Rigny
Photographer: Brisard Dampierre

项目地址：法国
竣工时间：2011 年
项目设计：DVVD
建筑用途：桥梁
开发商：Rigny 市政府
摄影师：Brisard Dampierre

Project Introduction

The River Saone divides Rigny in two parts. Essential for local journeys, a bridge which was built in 1904 allowed over 250 vehicles a day to cross. It was dismantled in 2009, after a detailed inspection and load-bearing study.

| | | remplissage garde corps | |
| mât | tête de pieux | treillis inox dans un cadre | lisse tube inox |

| PRS | | hauban | entretoise |
| 400 x 800 mm | micro pieux | PRS 350 x 350 mm | HEB 400 mm |

écorché structure acier

Structure Analysis Chart —— 结构分析图

1/100

项目介绍

索恩河将 Rigny 一分为二。该项目原建于1904年，曾是当地重要的桥梁，每日的车流量超过250辆。但在2009年经过仔细的检测和承重研究后，它被拆除了。

Design Conception

The award-winning project evokes the simplicity of the historic bridge. Contemporary design methods and improvements in materials have allowed the new structure to be slim and elegant. Instead of the two 2m-high lattice beams, today's 141.5 meters bridge is slimmer (the deck is 80cm thick) and graceful. Constructed using assembled steel profiles, it is painted white and rises gracefully from the existing stone pilings. At night, lighting enhances the linear nature of the deck and the rise of the masts skywards.

设计构思

该项目重现了旧时桥梁的简约之美,并获得奖项。现代的设计方法和材料上的改进,使得结构更轻便、更精致。舍弃了传统的两个 2 m 高的格构梁后,141.5 m 的桥体显得更加修长、雅致。组装后的钢材被固定在原有的石桩上,白色的丽影横跨水面。到了夜晚,灯光更是突出了桥面的流线感和直指苍穹的桥柱。

KURILPA BRIDGE
KURILPA 桥
THE ONLY SOLAR LIGHTING PEDESTRIAN OVERPASS
世界上唯一一座太阳能照明人行天桥

Location: Brisbane, Queensland, Australia
Completion year: 2009
Designer: Cox Rayner Architects
Use: Pedestrian Overpass
Client: Queensland Government Department of Public Works

项目地址：澳大利亚昆士兰州布里斯班
竣工时间：2009 年
项目设计：Cox Rayner 建筑事物所
建筑用途：人行天桥
开 发 商：昆士兰州市政工程处

Project Introduction

Kurilpa Bridge is the world's largest tensegrity bridge.It is not only a new pedestrian overpass, but also a new style public realm.It represents the status of Queensland in the area of art, science and technology. Kurilpa Bridge takes its name from the traditional owners of this stretch of the river who once used the area as a meeting place, walking along narrow riverbanks and pathways. Kurilpa translates as "place of the water rat". The bridge name was selected following a public naming competition and in consultation with the Turrbal and Jagera clans who lived in the area.

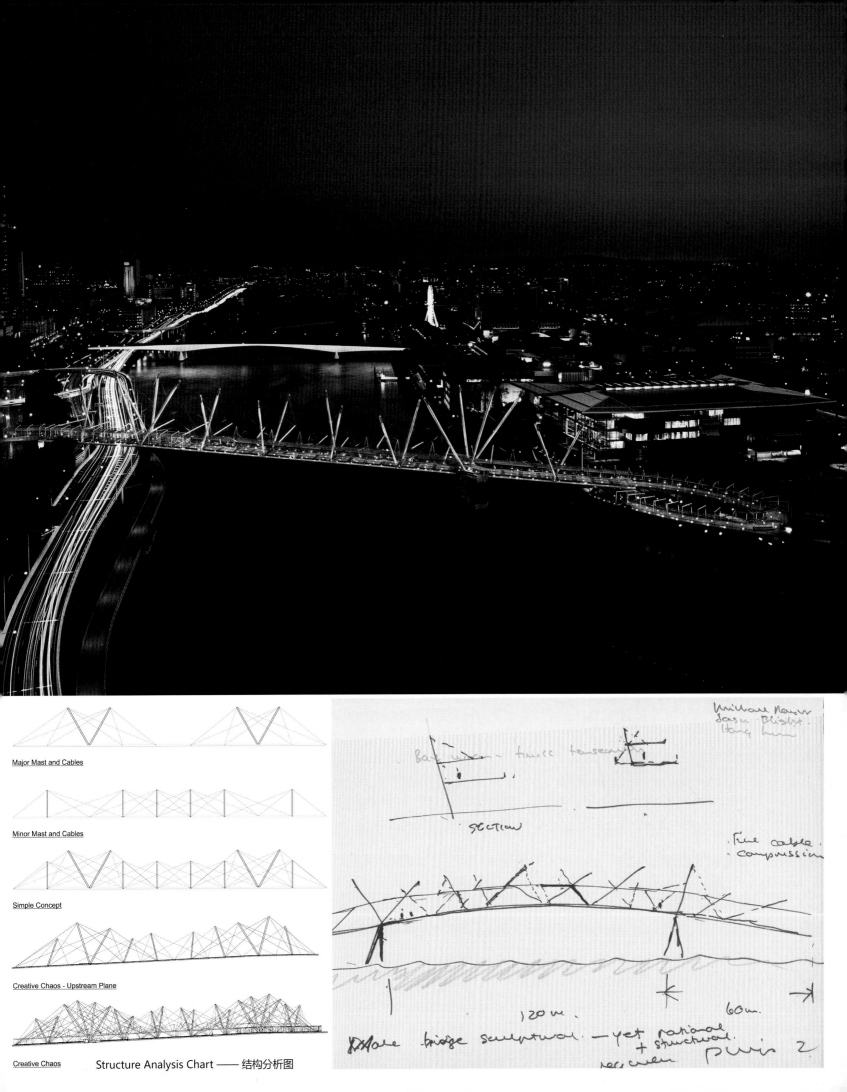

Major Mast and Cables

Minor Mast and Cables

Simple Concept

Creative Chaos - Upstream Plane

Creative Chaos

Structure Analysis Chart —— 结构分析图

项目介绍

该项目是世界上最大的张拉结构桥梁。它不仅仅是一个新的行人与城市交通的廊道，还是一个新形式的公共空间，体现出昆士兰在艺术、科学以及技术的前沿身份。库利尔帕步行桥因河岸一带的所有者曾将此地用作会议地点而得名，人们曾沿着狭窄的河畔人行道散步。库利尔帕步行桥又名"水老鼠的地盘"。步行桥的名字是在一次公众命名竞投中产生，并征得居住在当地的土著部落宗族的同意才定下来的。

Design Conception

A structural system with its origins in art and sculpture, "tensegrity", inspired Arup and Cox Rayner's design of the bridge. Sticks (compression masts and spars) and strings (cables) are arranged such that individual sticks do not touch. The result is a whimsical yet stable three-dimensional structure that appears to consist of a random array of floating spars and masts, presenting a superstructure that appears different from every viewpoint, providing an ever-changing visual experience for bridge users and the city in general. This structure not only creates a striking civic sculpture, but provides a highly efficient means of supporting an extraordinarily shallow deck as it sails over the Brisbane River and a busy expressway.

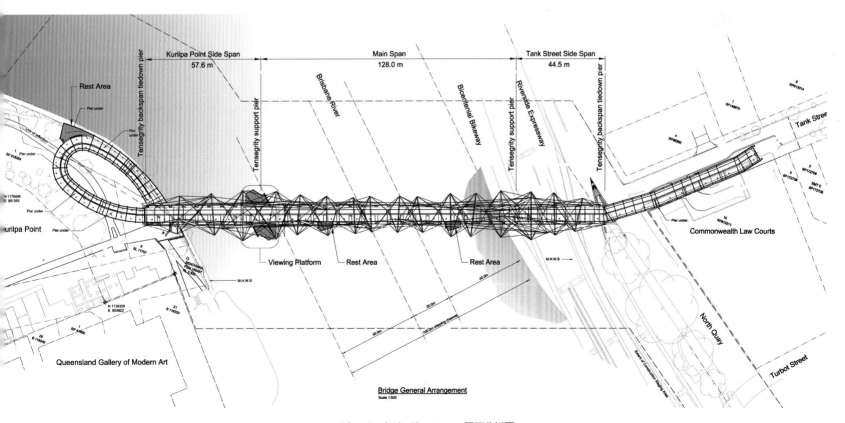

Plan Analysis Chart —— 平面分析图

设计构思

该项目的设计灵感来自艺术及雕塑领域的结构系统概念———张拉整体。多桅杆（由桅杆和圆管组成）和悬梁（缆绳）被排列成互不触碰的独立个体。其效果是一个奇妙但稳定的三维结构，仿佛由浮动的桅杆和圆管随意排列，组成上层架构。它从不同视角展现出不同的风采，为行人和整个城市带来了多变的视觉体验。这种结构不仅创造了一座引人注目的市政雕塑，还提供了一种高效的方式以支撑格外轻薄的桥面，跨越了里斯本河和一条繁忙的高速公路。

SUNWELL MUSE
SUNWELL 缪斯
GRACEFUL CRACK 优美的裂缝

Location: Tokyo, Japan
Completion year: 2008
Designer: Takatotamagami Architectural Design
Use: Show Room, Event Hall, Office
Site Area: 221m²
Client: Sunwell Group
Photographer: Masaya Yoshimura

项目地址：日本东京
竣工时间：2008 年
项目设计：Takatotamagami 建筑设计事务所
建筑用途：展厅、活动大厅及办公室
项目面积：221 m²
开 发 商：Sunwell 集团
摄 影 师：Masaya Yoshimura

Project Introduction

This is a building of a textile planning and trading company which handles the entire process from the production to retail. The site is located on the corner plot near the fashionable city "Harajuku".

Plan B1

Plan 1F

Plan 2F

Site Plan —— 平面图

项目介绍

Sunwell 是一家纺织与贸易公司,它掌控着产品从生产到销售的所有环节。这个知名的公司位于时尚的原宿区街角。

Plan 3F — Room-1

Plan 4F — Room-2

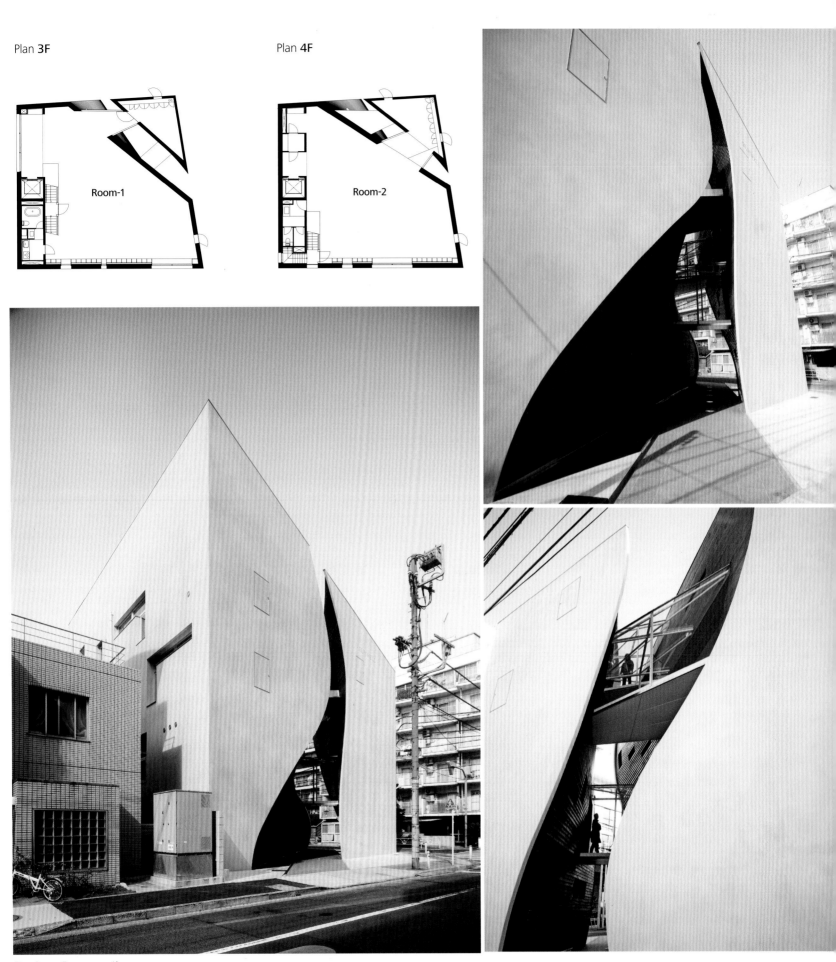

Design Conception

The client had been focusing on female apparel business, so the concept of our building design which is a metaphor of female beauty was suitable for them to put across their corporate identity. The components that characterize this architecture are the two curved surface walls which dominate the entire space. These two walls form a shortcut path which connects the roads in front of each side of the corner plot. This path looks like a narrow alley or the bottom of a ravine leads visitors inside the building, to the event hall in the basement and the showrooms on the first and second floors. The curves used in the elevation surfaces on the north side and the east side represents the beauty of a female body.

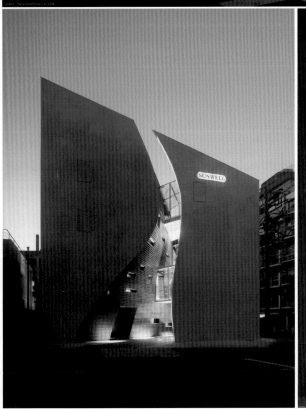

设计构思

因为 Sunwell 的主要业务是面向女性，所以设计团队需要在设计上展现女性的柔美，从而贴近企业的形象。两座弧状立面的墙体主导了整个空间，也正是该项目的独特之处。这两座墙构成了一条短途路径，连接两侧的公路。如窄巷又似沟壑的小径，将游客引入建筑内部，进入地下的活动大厅和一、二层的展览室。北面和东面的弧状裂缝展现出如女性线条般的美感。

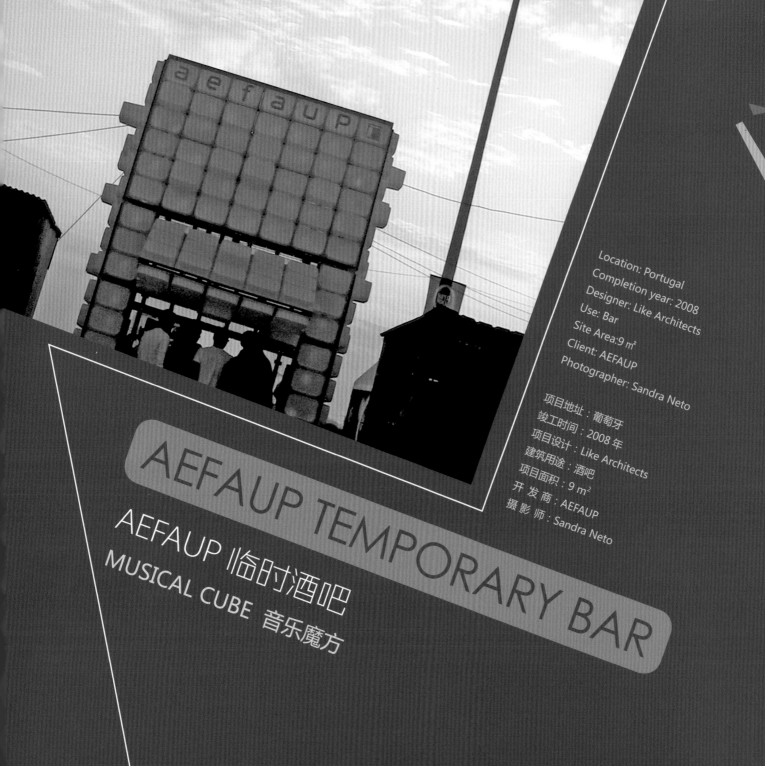

AEFAUP TEMPORARY BAR
AEFAUP 临时酒吧
MUSICAL CUBE 音乐魔方

Location: Portugal
Completion year: 2008
Designer: Like Architects
Use: Bar
Site Area: 9 m²
Client: AEFAUP
Photographer: Sandra Neto

项目地址：葡萄牙
竣工时间：2008 年
项目设计：Like Architects
建筑用途：酒吧
项目面积：9 m²
开 发 商：AEFAUP
摄 影 师：Sandra Neto

Project Introduction

This bar was a result of a competition aiming to represent the Faculty of Architecture in Porto. The given implantation, the fast construction and the low budget were some of the premises which had to be considered for designing this temporary structure that should be built in just one week with the help of students.

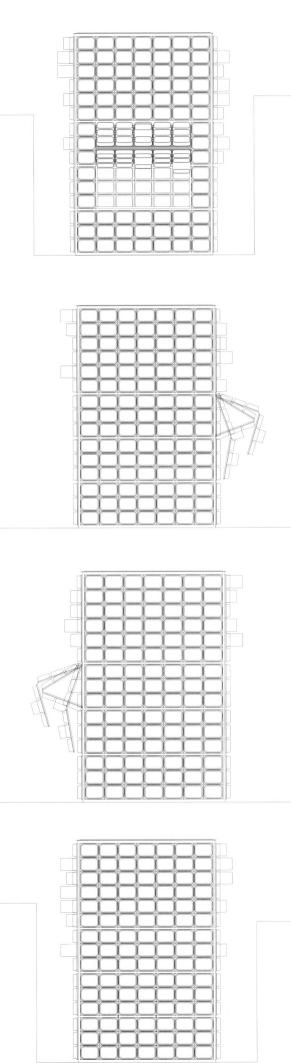

Elevation Drawing —— 立面图

项目介绍

该项目出自某比赛，体现了波尔图大学建筑系的设计水平。使用预制部件、短时间建成及低成本运作是该设计的前提。短短一周的时间，在同学们的帮助下，临时酒吧就在校园里搭建而成。

Design Conception

Departing from IKEA's concept "build-by-your-own", the project is a parallelepiped made out of different depth storage boxes resulting in a modular building with a textured skin, standing as a visual reference. The LED network in the concave interior spaces results in an exterior pattern at light, dramatically changing the bar appearance: at day a white abstract and closed volume; and at night the box will change it's light following the music.

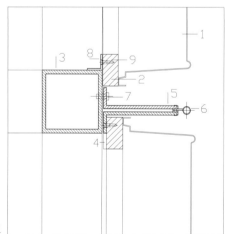

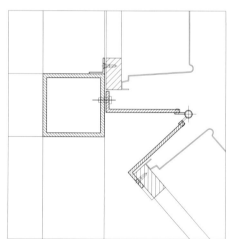

1 - Polyethylene plastic - storage boxes
2 - Secondary structure: wood stud 40mm
3 - Primary structure: stainless steel square bar 80/80mm
4 - Stainless steel flat bar 2mm
5 - Stainless steel angle 100/30/4mm
6 - Stainless steel hinges ø10mm
7 - Flat head sheet metal screw ø6mm
8 - Stainless steel angle 20/20/2mm
9 - Flat head wood screw ø4mm
10 - Stainless steel round bar ø12mm
11 - Stainless steel flat bar 4mm
12 - Stainless steel bent plate 40/40/4mm

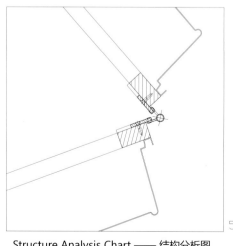

Structure Analysis Chart —— 结构分析图

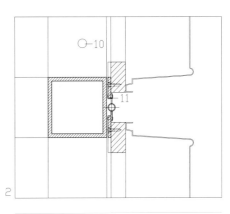
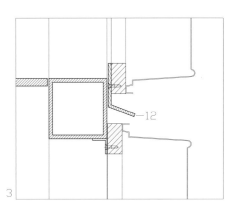

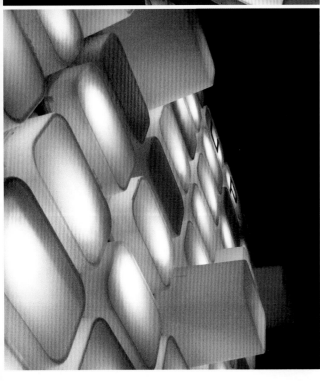
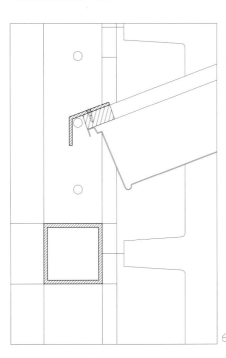

设计构思

从宜家"自己动手"的理念出发，设计团队运用了不同深度的塑料储物盒，筑成了这个模数法建筑。它呈平行六面体，表面特征明显，极具吸引力。在白天的时候，这座酒吧就是一个抽象而封闭的装饰品；到了晚上，箱子内的LED照明系统就会随着音乐的节奏闪烁起来，整个建筑的外观也随之变化。

Location: New York, USA
Completion year: 2011
Designer: HWKN
Use: Showcase
Client: UNIQLO
Photographer: Michael Moran

项目地址：美国纽约
竣工时间：2011 年
项目设计：HWKN
建筑用途：商品展示柜
开 发 商：UNIQLO
摄 影 师：Michael Moran

UNIQLO CUBES
优衣库方块
WHITE MAGIC BOX 白色魔方

Project Introduction

The cubes are equipped with glowing façades and integrated shelving structures, simple and striking at once, to host UNIQLO's unique products. The pop-up stores show case UNIQLO by forming simple volumes with high-tech surfaces cladding their gridded forms.

项目介绍

优衣库委托 HWKN，设计了两个小巧的方块体展示装置。它们拥有闪亮的外表，极具吸引力。另外，其内也配备了综合陈列设施，用于展示出色的优衣库产品。高科技电镀技术使其网格表面更具魅力，配合特别设计的弹出式衣橱，不乏优衣库的时尚风格。

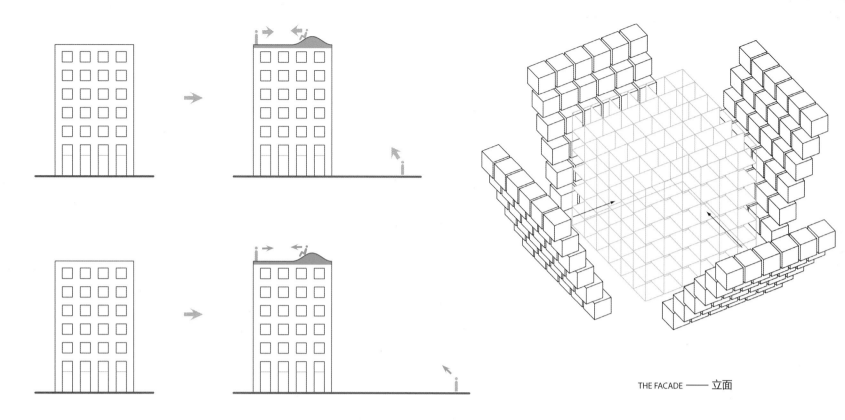

THE FACADE —— 立面

Design Conception

A section of each cube slides open like a vault, inviting people to come visit, shop the collections, and purchase the treasures that lay inside. At night all cube glow in the dark to mark their presence. UNIQLO's brand presence emerges from the powerful simplicity of the glowing forms and at the high level of material quality and detailing.

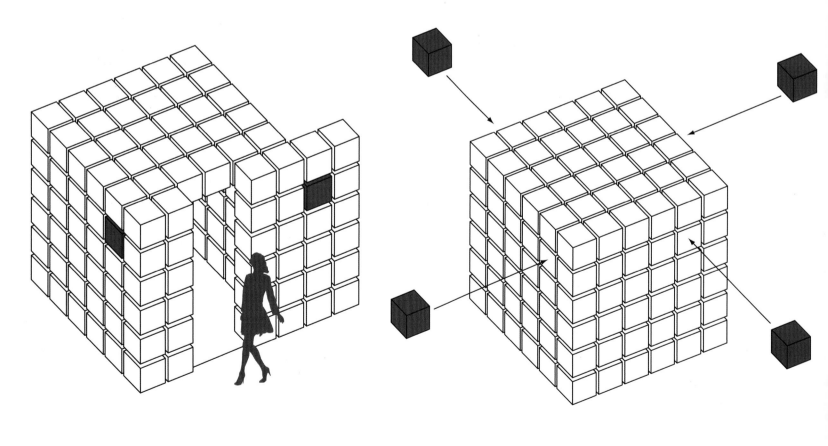

THE ENTRANCE —— 入口

THE UNIQLO CUBE —— 优衣库方块

设计构思

设计师匠心独运,使方块上的一部分能随时像门一样滑开,仿佛一下子将客人带进神秘、美妙的世界里。对他们来说,陈列着的产品就如同正待挖掘的宝藏。夜晚时,方块如萤火虫般闪烁,在漆黑的广场上熠熠生辉。如此这般,优衣库的品牌特性得以凸显,征服大众。

SPORTS-PAVILION
鹿特丹体育俱乐部
LIGHTWEIGHT BUILDING 简约轻便

Location: Rotterdam, The Netherlands
Completion year: 2010
Designer: Erik Moederscheim
Use: Football Club
Site Area: 1,410 m²
Client: Municipality of Rotterdam
Photographer: Rob't Hart

项目地址：荷兰鹿特丹
竣工时间：2010 年
建筑设计：Erik Moederscheim
建筑用途：足球俱乐部
项目面积：1 410 m²
开 发 商：鹿特丹市政府
摄 影 师：Rob't Hart

Project Introduction

The project is part of the development of Park 16Hoven; a large new suburban neighborhood adjacent between the city center and the airport.

项目介绍

该项目是 16Hoven 公园开发项目的一部分，位于市中心和机场之间新的郊区住宅区域。

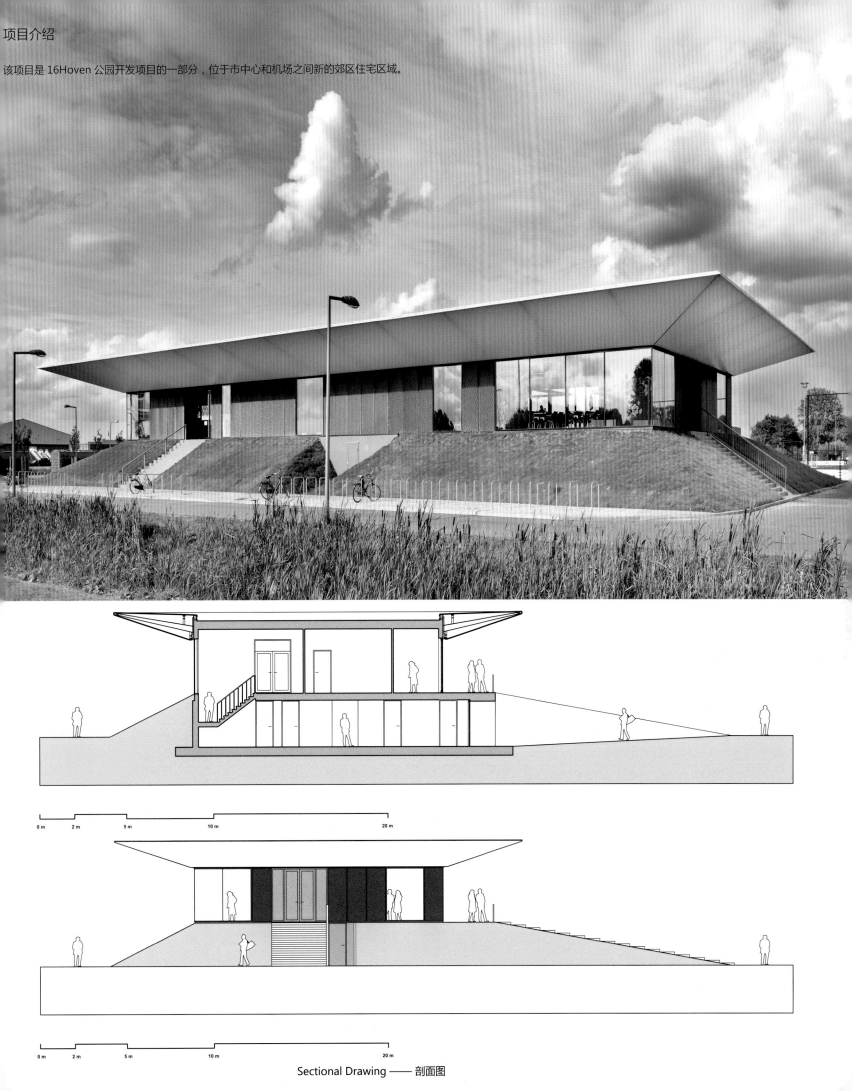

Sectional Drawing —— 剖面图

Ground Floor Plan —— 地面平面图

First Floor Plan —— 一层平面图

Design Conception

Within the open space between the airport and newly developed houses in the park, the aim for the design was to create a transparent and 'lightweight' pavilion. The building is set up in two levels. The clubhouses and the boardrooms are situated on the top floor of the building. This level is directly connected to the pitches by grass-covered slopes. These slopes cover the ground floor with its dressing rooms and storage areas, and provide a natural grandstand for audience. The final piece of the building is the translucent cantilevered roof. This roof filters direct sunlight and illuminates like a lampoon in the evening thanks to the integration of LED powered lighting.

设计构思

设计师的目标是在机场和新建的住宅群之间的开放空间里，建立一个具有透明和轻便特点的体育运动馆。该项目是一个两层建筑。会所和会议室位于顶层。顶层直接连接草地覆盖的斜坡球场。这些斜坡覆盖范围包括了更衣室和地下储存区，同时为观众提供了一个天然看台。体育馆的顶部是半透明的悬臂式屋顶。屋顶白天可以过滤直射阳光，晚上依靠 LED 照明设备烘托轮廓。

HAPPY STREET
快乐街
NETHERLANDS PAVILION 上海世博会荷兰馆

Location: Shanghai, China
Completion year: 2010
Designer: Rijk Blok
Use: Exhibition
Site Area: 5,000 m²
Client: Ministry of Economic Affairs

项目地址：中国上海
竣工时间：2010 年
项目设计：Rijk Blok
建筑用途：展览
项目面积：5 000 m²
开 发 商：Ministry of Economic Affairs

Project Introduction

The 5,000-square-meter Netherlands Pavilion in Zone C is themed "Happy Street." The Netherlands Pavilion resembles the figure eight, a lucky number suggesting fortune in Chinese. The up-and-down pavilion consists of 26 small houses along a curving pedestrian walkway.

项目介绍

荷兰馆的主题为"快乐街",位于上海世博园C区。街道呈"8"字形,在中国有吉利的寓意。起伏的展馆包括26间小房子,沿着弯曲的步道分布开来。

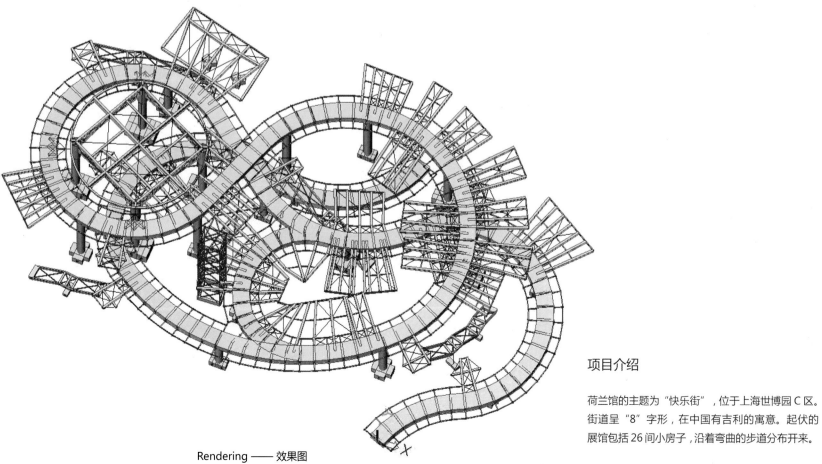

Rendering —— 效果图

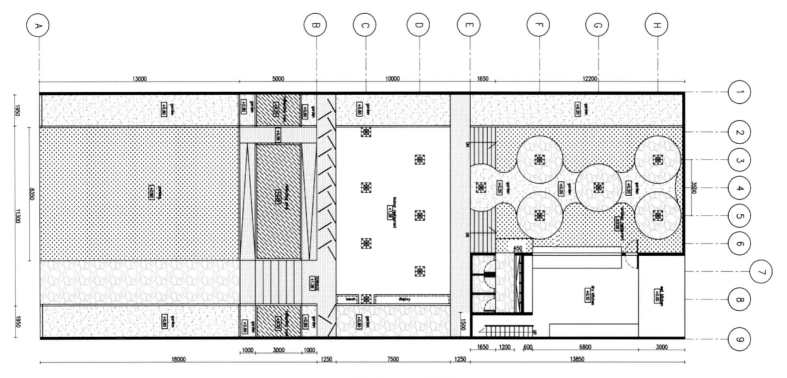

Site Plan —— 平面图

Design Conception

The form of restaurant umbrelta basically was inspired by an umbrella. But we want to make a "giant" and an "unusual" umbrella. The "giant" form would be functioned as a "shelter". A group of "shelter" then becomes a "building", which hopefully could protect the restaurant's guest from sun, wind, and rain. We make the "giant" bamboo umbrella in different size (the width and the height) for arranging the "giant" umbrella in such a way that each "umbrella" could overlap each other to become a "giant" roof. The "giant" roof functioned as a shelter and also as a gutter for rain. The rain water is directly distributed to the ground via a pipe in the middle of the structure, which is wrapped by Bamboo. Bamboo is a strong material and with a minor modification, we can directly create a great and beautiful structure.

设计构思

建筑形式的设计最初是受到雨伞的启发。伞大而与众不同。因为"大",它可以起到遮挡的作用。由竹伞构成的建筑,可以为就餐的顾客遮挡阳光、抵御风雨。竹伞的宽度和大小不一,互相重叠,形成了巨大的屋顶。这样的结构还起到了排水的作用。雨水通过中间的竹管,直接排到地面。竹子是非常坚韧的材料,而且只需小小的修改,便可以构造出大气而美观的建筑。

747 WING HOUSE
747机翼屋
WHEN THE PLANE CHANGE INTO A HOUSE 当飞机走进住宅

Location: Malibu, CA, USA
Completion year: 2011
Designer: Studioea
Use: Residence

项目地址：美国加利福尼亚马里布
竣工时间：2011 年
项目设计：Studioea
建筑用途：住宅

Project Introduction

This project exists on a 55-acre property in the remote hills of Malibu with unique topography and panoramic views looking out to a nearby mountain range, a valley, and the Pacific Ocean with islands in the distance.

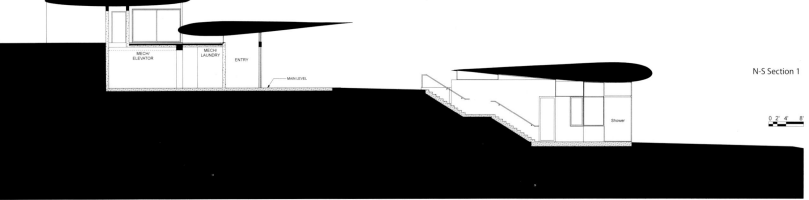

Sectional Drawing —— 剖面图

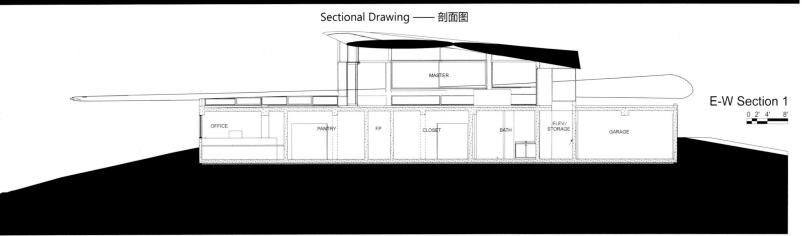

Elevation Drawing —— 立面图

项目介绍

该项目位于马里布偏远山村,一块 55 英亩的地块上。这里拥有独一无二地形特征。在这能够看到周边山麓的全景,近处的山谷及远处太平洋上的岛屿。

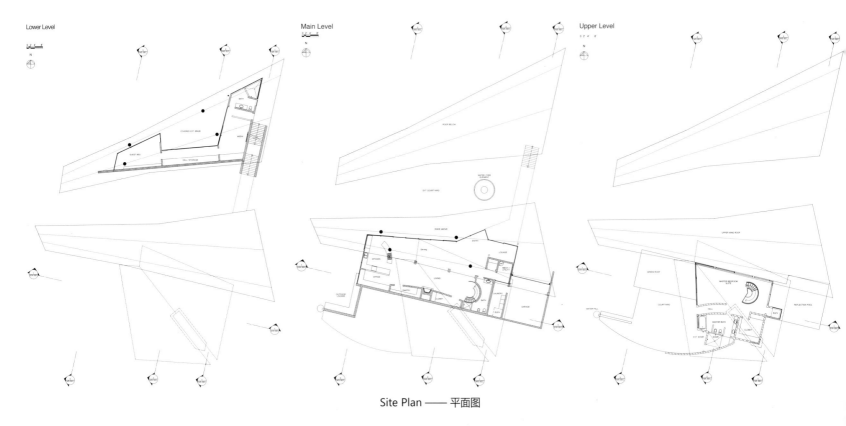

Site Plan —— 平面图

Design Conception

The master bedroom has used both of the main wings as well as the 2 stabilizers from the tail section as a roof for the Master Bedroom. The Art Studio Building has used a 50-foot long section of the upper fuselage as a roof, while the remaining front portion of the fuselage and upper first class cabin deck will be used as the roof of the guest house. The lower half of the fuselage, which forms the cargo hold, will form the roof of the Animal Barn. A Meditation Pavilion has been made from the entire front of the airplane at 28 feet in diameter and 45 feet tall; the cockpit windows formed a skylight. Several other components are contemplated for use in a sublime manner, which include a fire pit and water element constructed out of the engine cowling.

设计构思

主卧室的两侧插上了两个机翼,在主卧的盥洗室后安置了一个水平尾翼。室内的艺术工作室区域采用了 50 英尺长前部机身作为调高的天花,而余下的机身头等舱部分则用作会客室的天花。机身的下半部改造成了储藏室,用于储藏谷物以及饲养家畜。临时展馆 Meditation Pavilion 则是由飞机的整个前片制造而成的,建造后的尺寸直径足有 28 英尺、高 45 英尺。为了室内通风透光,飞机驾驶舱的窗口被改造成了天窗。飞机上的其他组件也被全部利用了起来,譬如天然气通路及自来水管等,都是这些零部件制造的。

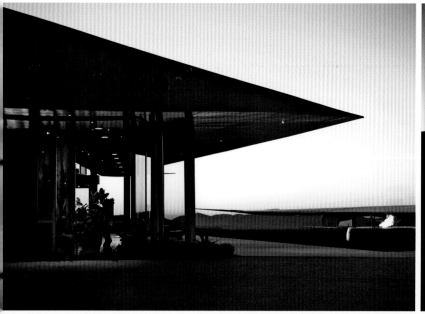

COCOON
蚕茧大厦
EXCLUSIVE OFFICE HEADQUARTER, ZURICH 苏黎世高级总部办公楼

Location: Zürich, Switzerland
Completion year: 2007
Designer: Camenzind Evolution AG
Use: Office
Site Area: 1,900 ㎡
Client: Swiss Life
Photographer: Camenzind Evolution

项目地址：瑞士苏黎世
竣工时间：2007 年
项目设计：Camenzind Evolution AG
建筑用途：办公
项目面积：1 900 m²
开 发 商：Swiss Life
摄 影 师：Camenzind Evolution

Project Introduction

Cocoon is located in Zurich's Seefeld district on a beautiful hillside, which enjoys excellent lake and mountain views. The location's distinctive flair stems from the exceptional park-like setting – a green oasis into which Cocoon snugly nestles. Flanked on three sides by mighty, old trees the elliptical structure reads as a freestanding sculptural volume that gracefully rises up from the park.

项目介绍

COCOON 总部办事处位于苏黎世塞费尔德地区的一座美丽的山坡上，拥有极好的山湖自然景观。所处地段极具标志性，类似公园的配置——办公楼依偎在绿洲之中。椭圆形的结构拥有一个独立的、雕塑般的体量，它三面被古树包围，优雅而从容地从花园中盘旋升起。

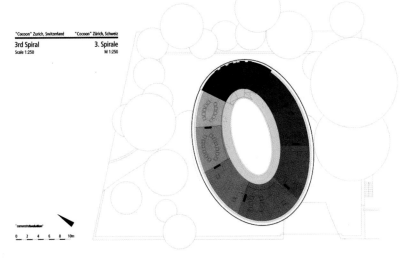

Site Plan —— 平面图

Design Conception

The bold stand-alone building embodies an innovative conception of interior spatial organization and interaction with the surrounding environment. With its spiral massing, Cocoon may be conceived as a sort of "communication landscape" that creates a unique spatial configuration and working environment in a matchless setting. The stepped, upward-winding sequence of segments also shapes the character of the building interior. All spaces are arranged along a gently rising ramp, which wraps around a central, light-flooded atrium. The space planning concept dispenses with the traditional division into horizontal storeys in favour of a seemingly endless sequence of elliptical floor segments.

设计构思

这个大胆而独立的建筑为内部空间组织以及与周围环境互动提出了创新理念。螺旋形的体量赋予建筑"沟通景观"的涵义，创造了一个无与伦比的空间结构和工作环境。阶梯在向上环绕的同时塑造了内部空间的特征。所有的内部空间都围绕着建筑中央的采光中庭，并沿着缓缓上升的楼梯安置。其楼层规划摒弃了传统的划分方式，采用一个个椭圆形隔断，产生了无止境的视觉效果。

GATE 750
大门 750
LOGO OF THE FESTIVAL 节日标志

Location: Holland
Completion year: 2008
Designer: Mulders vandenBerk Architecten
Use: Jubilee Pavilion 750th anniversary Amersfoort
Photographer: Roel Backaert

项目地址：荷兰
竣工时间：2008 年
项目设计：Mulders vandenBerk Architecten
建筑用途：纪念阿默斯福特市 750 周年华诞
摄 影 师：Roel Backaert

Project Introduction

A clearly identifiable and multipurpose building was requested to organize the 750th anniversary of the city Amersfoort. The shape of the sculpture and the façade print give Gate 750 a strong identity. This makes the building becoming the logo of the festival.

项目介绍

阿默斯福特市为了庆祝750周年华诞,要求设计团队设计一个多功能、具有标示性的建筑。建筑的外形与立面极具个性,使其很快成为了节日的标志。

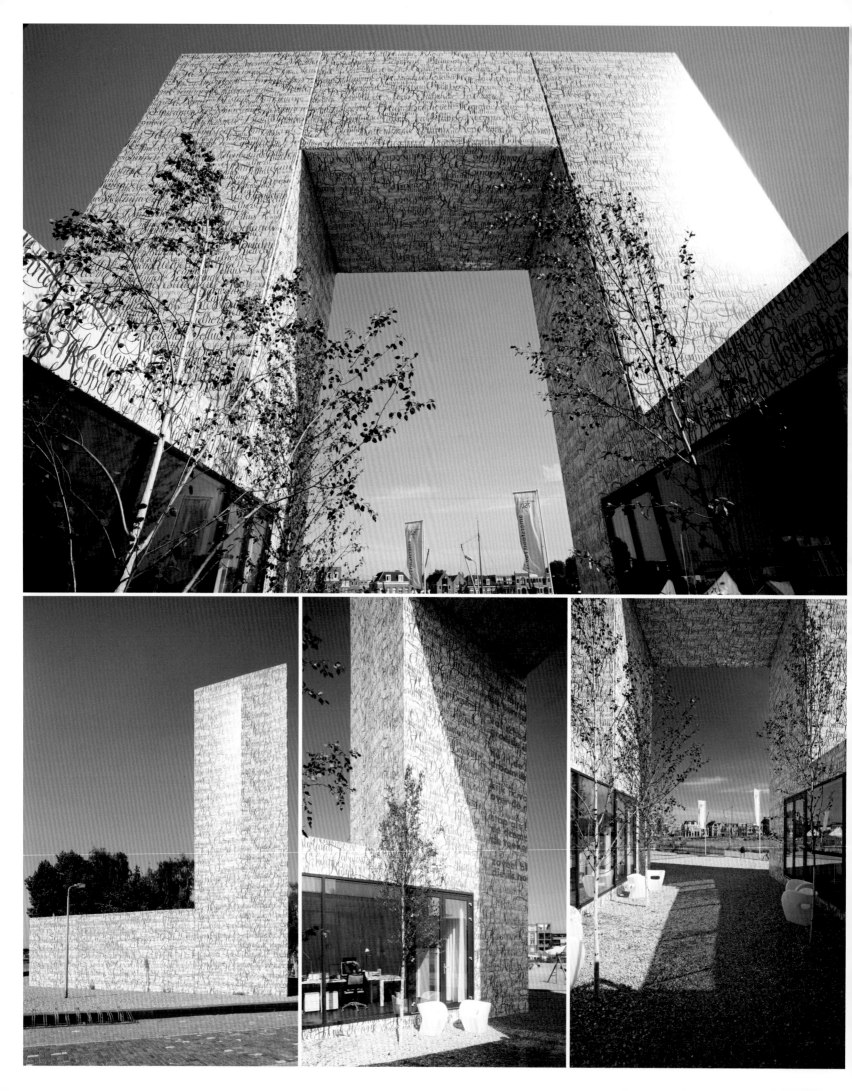

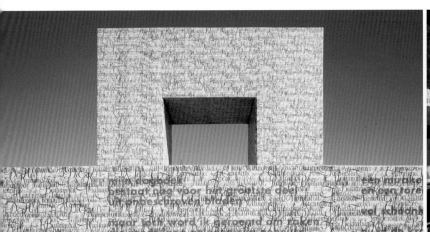

Design Conception

The entrance is formed by a central patio, surrounded by meeting, exhibition and office space. This spatial setting creates a good overview and interaction within the building. The spaces can be used separately or combined as one continuous space for different uses. The gate can be entered to get a viewing platform on top. From this platform one can overlook the festival area. The whole building is wrapped in fabric on which words and a poem about Amersfoort is printed.

设计构思

整个入口由中心天井及周围的会展、会议室、办公空间组成。这样的空间结构，为室内带来了很好的视野和交流。这些空间既可以分开利用，也可以合并成一个多功能连续空间。大门的顶部设有观景平台。在平台上，你可以俯瞰整个节日现场。建筑表层全都包裹着印有歌颂阿默斯福特市诗词的织物。

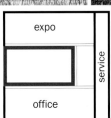

by separating office and expo, an extra space is created; the patio

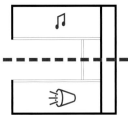

the separation makes a diverse use possible

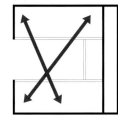

by making visualy one space, there is a good overview

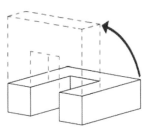

folding the volume

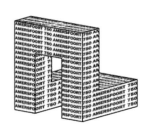

apearance, interaction, public function

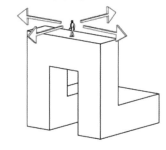

view to surrounding

Design Analysis Chart —— 设计分析图

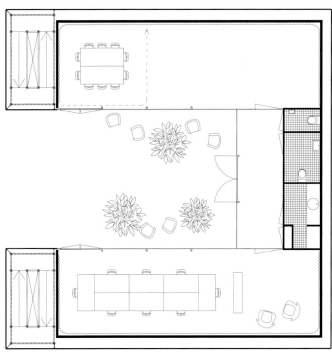

Site Plan —— 平面图

ISERNIA GOLF CLUB
伊塞尔尼亚高尔夫俱乐部
ADDING ICONIC VALUE 加强标示性

Location: Isernia, Italy
Completion year: 2010
Designer: Medir Architects
Use: Club
Site Area: 260 m²
Client: STM Srl
Photographer: Courtesy of Medir Architects

项目地址：意大利伊塞尔尼亚
竣工时间：2010 年
项目设计：Medir Architects
建筑用途：俱乐部
项目面积：260 m²
开 发 商：STM Srl
摄 影 师：Courtesy of Medir Architects

Project Introduction

The Isernia Golf Club will be soon part of a new luxury holiday centre that is being built within a wood, located just near the golf course.

项目介绍

该项目将在近期内成为新豪华度假中心的一部分，全木制，邻近高尔夫球场。

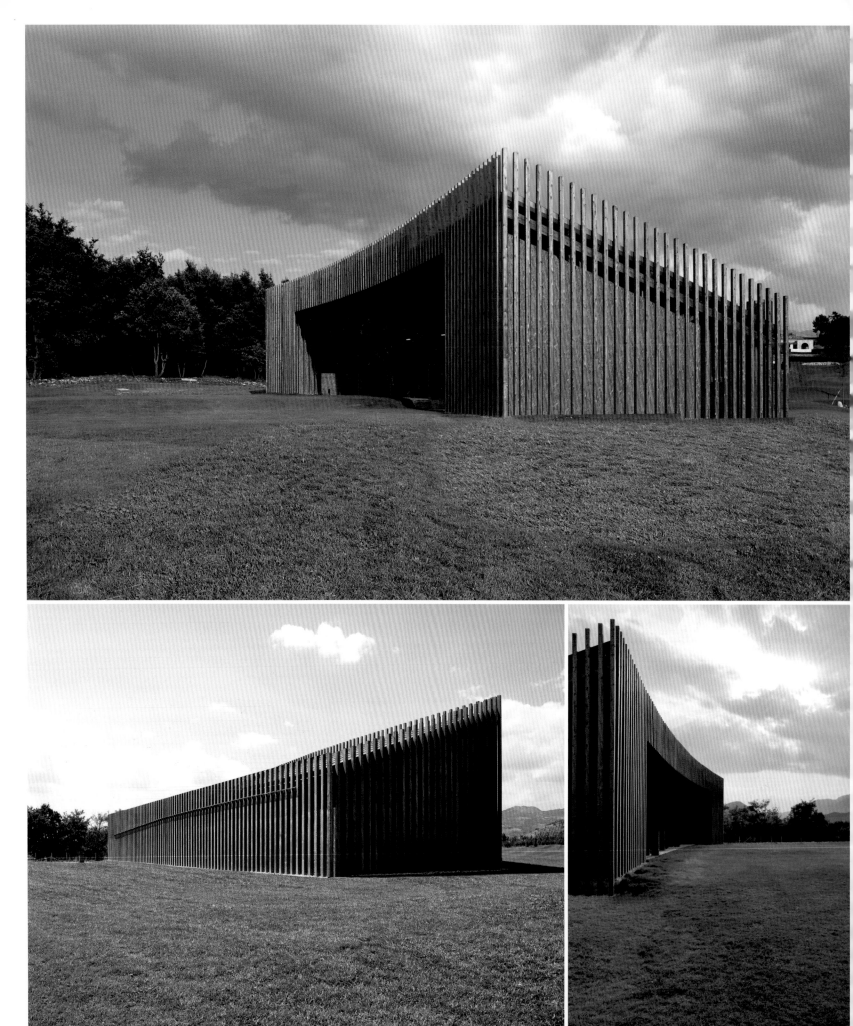

Design Conception

The design concept of the Isernia Golf Club building started with the idea of converting an existing wooden structure, built it to cover the golf shooting stations, into a Clubhouse pavilion. The original structure of that shelter was static and got lost with generous size of the golf course and the beautiful open-ended views facing the surrounding valleys. Therefore, the intention was to enhance the presence of the building by adding iconic value to the imperfections of its shaped geometry.

设计构思

这个高尔夫俱乐部的设计理念，源于一个已有木质结构的改造，即建造一个用于覆盖高尔夫发球台的俱乐部亭子。原有的结构过于死板，已经被巨大的球场和周围美丽的景色掩盖了。因此设计的目的在于改变原来的几何外形，加强建筑的标志性。

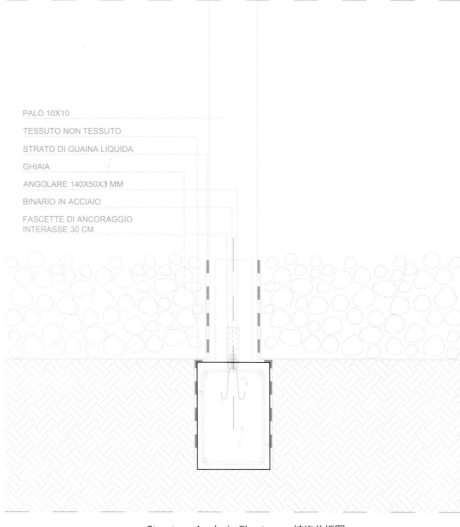

PALO 10X10
TESSUTO NON TESSUTO
STRATO DI GUAINA LIQUIDA
GHIAIA
ANGOLARE 140X50X3 MM
BINARIO IN ACCIAIO
FASCETTE DI ANCORAGGIO INTERASSE 30 CM

Structure Analysis Chart —— 结构分析图

Panoramic View Plan —— 全景平面图

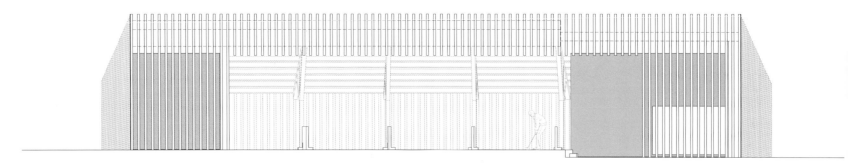

CLUBHOUSE, PROSPETTO NORD

Elevation Drawing —— 立面图

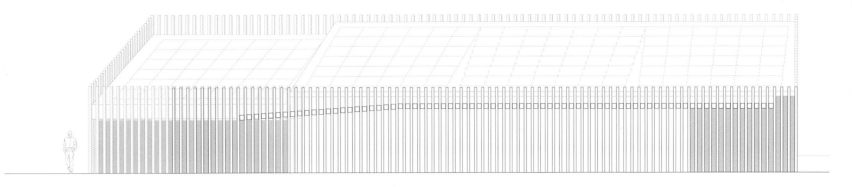

CLUBHOUSE, PROSPETTO SUD

Elevation Drawing —— 立面图

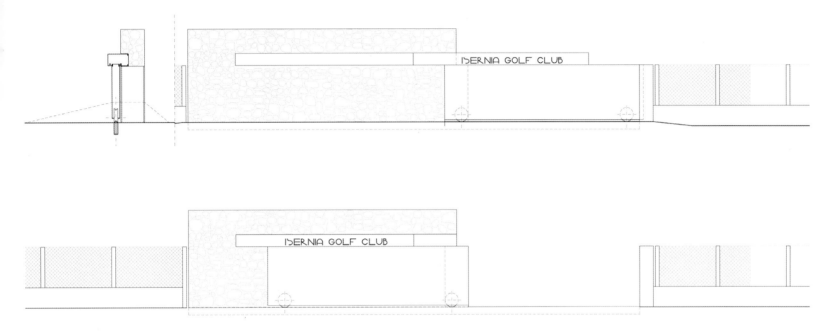

Elevation Drawing —— 立面图

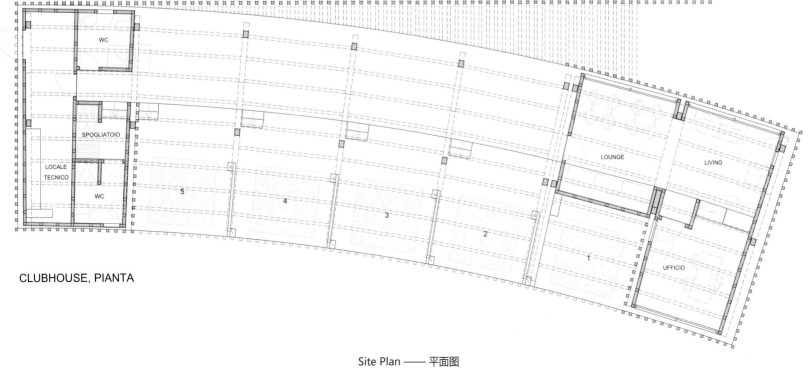

CLUBHOUSE, PIANTA

Site Plan —— 平面图

KOBAN
派出所岗亭
AN UNUSUAL POLICE OFFICE 颠覆传统

Location: Kumamoto, Japan
Completion year: 2011
Designer: Klein Dytham architecture
Use: Police Box
Site Area: 143.62 m²
Client: Kumamoto Prefectural Police Headquarter

项目地址：日本熊本县
竣工时间：2011年
项目设计：Klein Dytham architecture
建筑用途：警察岗亭
项目面积：143.62 m²
开 发 商：Kumamoto Prefectural Police Headquarters

Project Introduction

KDa's new koban, a neighborhood police station, is located near Kumamoto's new railway station and was required to stand out as a local landmark but present a friendly image. The building stands on a teardrop-shaped site, separated from the large surrounding buildings by taxi station and a tramline.

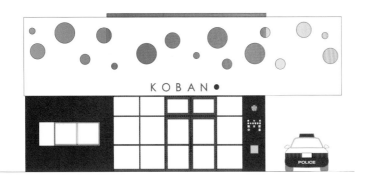
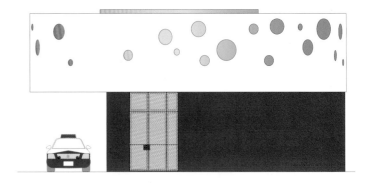
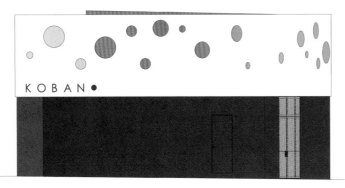

Elevation Drawing —— 立面图

项目介绍

因为这个街道派出所毗邻熊本县新建的火车站，所以设计团队被要求将其打造成一个外观亲切的地标式建筑。该项目坐落于一个泪滴形状的地块上，通过的士站点和电车轨道与周围大体量的建筑分离。

Design Conception

To make the requested sculptural gesture, KDa wrapped the top of the building in a ribbon of perforated steel plate and colored the upper floor volume and the inside of the ribbon with a gentle rainbow gradation. This bold graphic can be seen from all around by the neighborhood, but up close creates subtle effects by casting shadows on the surrounding road. While the ribbon was not part of the brief for the building, KDa have cunningly made it functional - a 3 m cantilever creates a shelter where patrol cars can park, allowing people and exit without getting wet!

设计构思

为了达到雕塑般的装饰效果，设计团队为建筑的顶部包裹了一圈缎带状的穿孔钢板，并将上层的体量涂色，还在"缎带"的内侧刷上彩虹般的渐变色涂料。周围的居民区都可以看到这个创意十足的装饰性建筑。而近处的马路也因带孔的"缎带"变得光影斑驳。同时，屋顶向外悬挑 3 m，形成下面的停车空间，能为人们遮风挡雨。

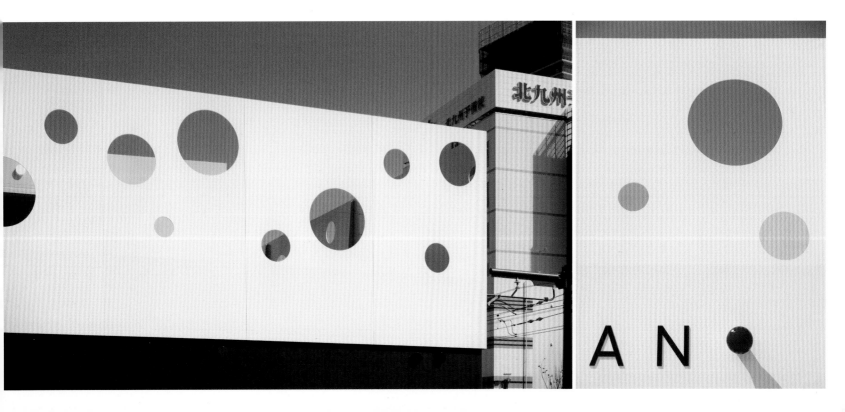

KALMAR MUSEUM OF ART
卡尔马艺术博物馆
BLACK CUBE 黑色方块

Location: Kalmar, Sweden
Completion year: 2008
Designer: THAM & VIDEGÅRD ARKITEKTER
Use: Exhibition
Site Area: 1,594 m²
Client: Municipality of Kalmar
Photographer: Åke E'son Lindman

项目地址：瑞典卡马尔
竣工时间：2008 年
项目设计：THAM & VIDEGÅRD ARKITEKTER
建筑用途：展览
项目面积：1 594 m²
开 发 商：卡马尔市政府
摄 影 师：Åke E'son Lindman

Project Introduction

The Kalmar Museum of Art is the result of a winning proposal in the open international competition in 2004 and was inaugurated on the 10th of May 2008. Set among the high trees in the city park of the renaissance town of Kalmar, it is built on part of the remains of the medieval city wall, next to a restaurant pavilion dating from the 1930s by Swedish modernist Sven Ivar Lind.

卡尔马艺术博物馆是根据2004年开放式国际设计大赛的一套获奖方案建造而成,并于2008年5月10日正式对公众开放。博物馆坐落于文艺复兴城市卡尔马的城市公园的树丛中,毗邻一座20世纪30年代建成的餐厅,这家餐厅是瑞典现代主义建筑大师斯文伊瓦尔·林德(Sven Ivar Lind)设计的作品。

Panoramic View Plan —— 全景平面图

项目介绍

卡尔马艺术博物馆是根据2004年开放式国际设计大赛的一套获奖方案建造而成,并于2008年5月10日正式对公众开放。博物馆坐落于文艺复兴城市卡尔马的城市公园的树丛中,毗邻一座20世纪30年代建成的餐厅,这家餐厅是瑞典现代主义建筑大师斯文伊瓦尔·林德(Sven Ivar Lind)设计的作品。

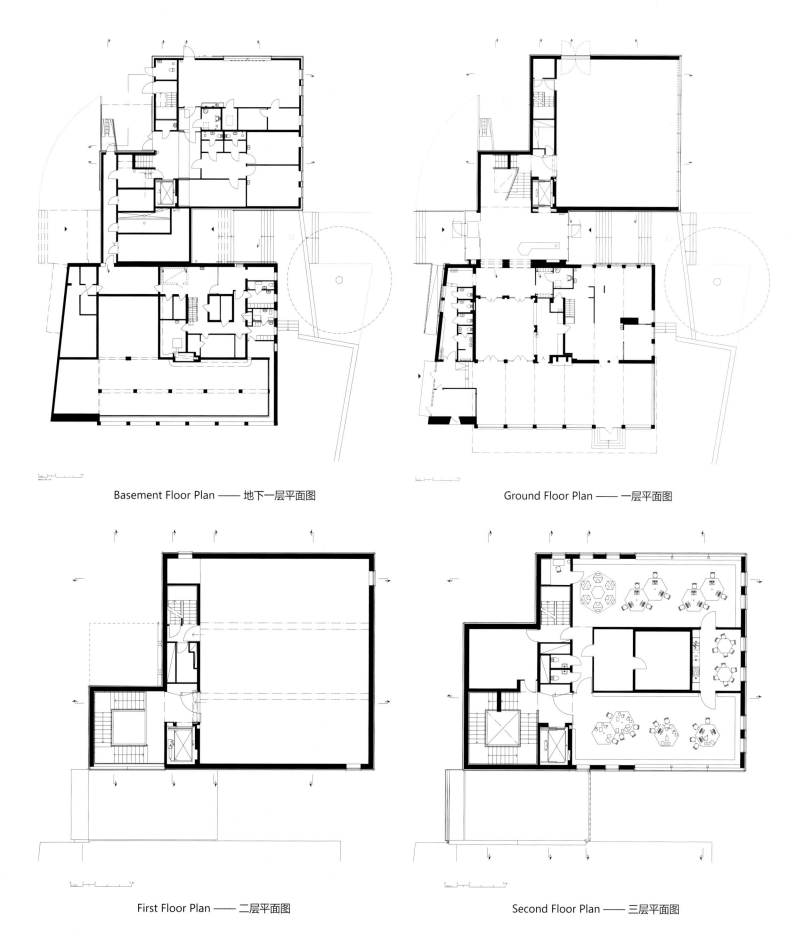

Basement Floor Plan —— 地下一层平面图

Ground Floor Plan —— 一层平面图

First Floor Plan —— 二层平面图

Second Floor Plan —— 三层平面图

Design Conception

The new museum is a black four-level cube clad with large scale wooden panels and punctuated by large glazed openings. The two main spaces are the white boxes where one side can open up completely to bring in the exterior of the park. One of the architectural main features is the open stair spiralling the full height of the building, starting from the new entrance lobby that interconnects between lake-side and park. It is a top lit space with all surfaces finished in exposed in situ cast concrete. The four floors, each is different from the others, stacked on top of each other and create a vertical walk up into the greenery of the trees with a series of different spatial experiences while offering views of the environs; the Kalmar castle, the lake and the city centre.

设计构思

新博物馆是一座4层的黑色方块式建筑,外覆大型木质板材,搭配巨大玻璃窗体。两块形如白色的盒子,一侧可完全打开,与外部公园环境连通。建筑的主要特点之一就是盘旋于整个楼体上的开放式楼梯。而大厅作为楼梯的起点实现了湖岸一侧与公园的互动,空间采光极佳,所有现浇混凝土表面均直接裸露在外。四层楼面各有不同,相互堆叠,在树木绿化环境中拔地而起,实现了不同的空间体验,同时提供了卡尔马城堡、湖区以及市中心等周边环境的景观视野。

MIRROR HOUSE
镜宅
DISTORTING MIRROR 孩子们的哈哈镜

Location: Copenhagen City, Denmark
Completion year: 2011
Designer: MLRP Architecture
Use: Restroom
Site Area: 140 m²
Client: City of Copenhagen (CAU)
Photographer: Stamers Kontor

项目地址：丹麦哥本哈根市
竣工时间：2011 年
项目设计：MLRP 建筑事务所
建筑用途：洗手间
项目面积：140 m²
开 发 商：哥本哈根市政
摄 影 师：Stamers Kontor

Project Introduction

The Mirror House is a flexible space and restrooms, used by kindergarden classes.

Panoramic View Plan —— 全景平面图

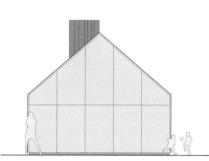
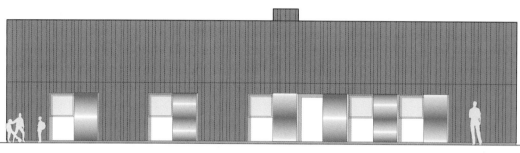

SOUTH-EAST Sectional Drawing —— 剖面图 NORTH-EAST

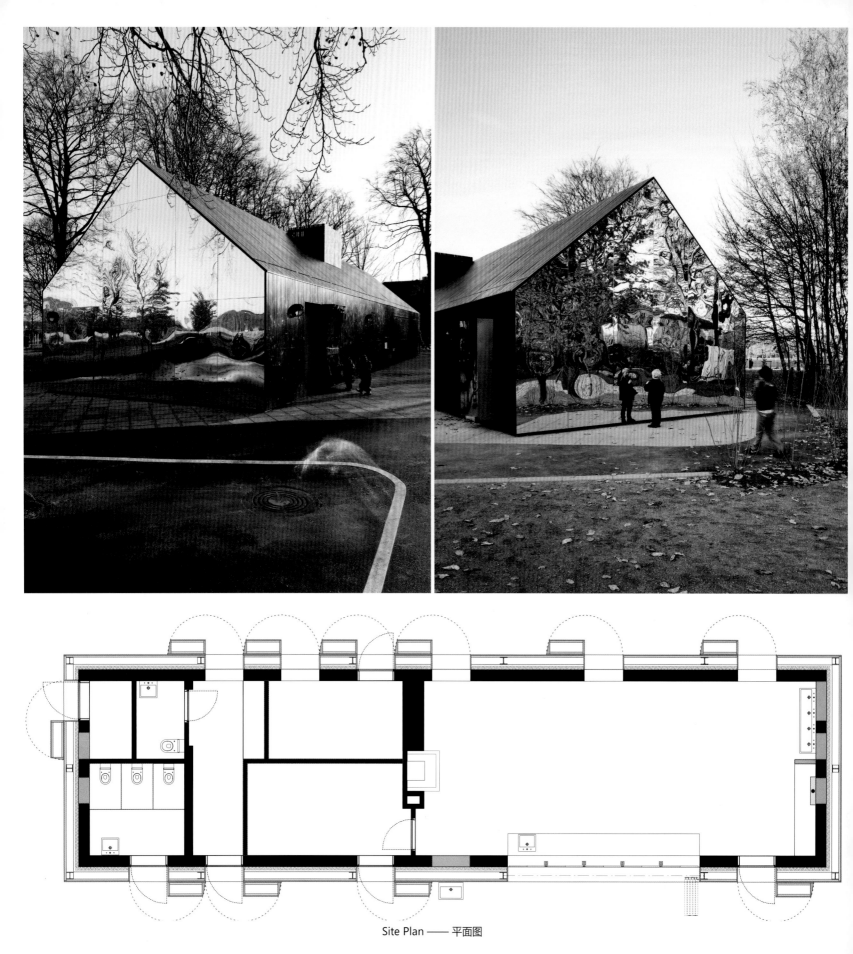

Site Plan —— 平面图

项目介绍

该项目为幼儿园的孩子们提供了灵活的空间及多个休息室。

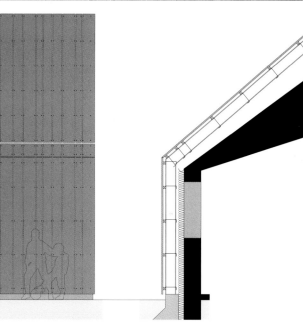

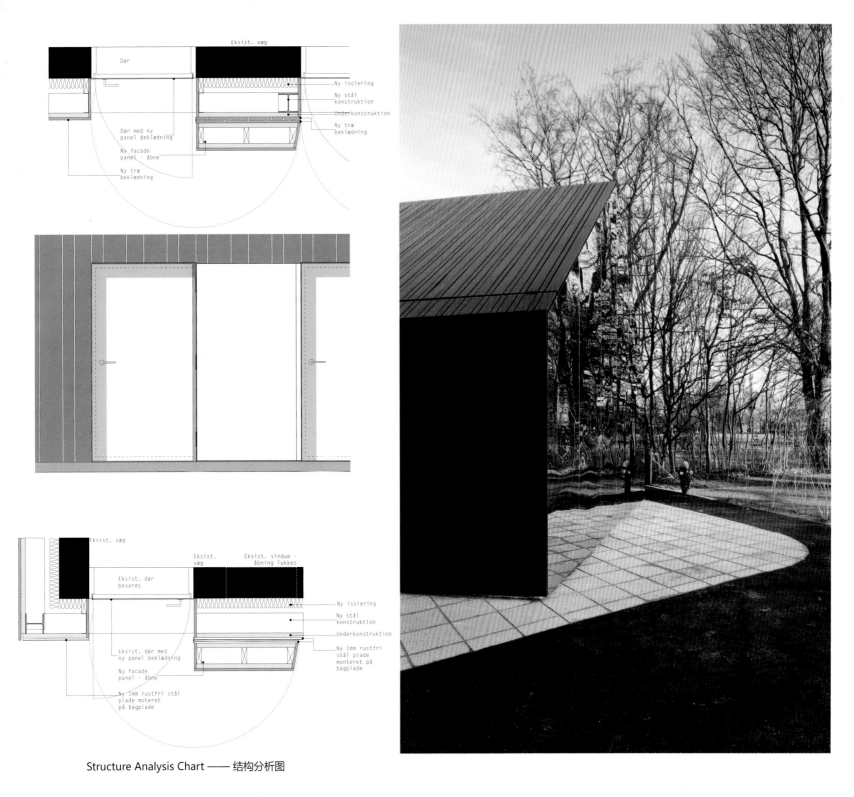

Structure Analysis Chart —— 结构分析图

Design Conception

Funhouse mirrors are mounted on the gabled ends of this playground pavilion in Copenhagen, as well as behind the doors. Instead of a typical closed gable facade, the mirrored gables creates a sympathetic transition between built and landscape and reflects the surrounding park, playground and activity. Windows and doors are integrated in the wood-clad facade behind facade shutters with varied bent mirror panel effects. At night the shutters are closed and become anonymous. During the day the building opens up, attracting the children who enjoy seeing themselves transformed in all directions.

设计构思

建筑的山墙以及门背面都装有镜子，通过反射，带来了有趣的视觉体验，让古板封闭的山墙变成"镜之墙"，并在建筑与景观之间巧妙的过渡，反映出园内的环境和活动。门与窗集中在木制百叶结构中，背面装有弯曲镜面。夜晚关闭时，建筑简洁隐蔽；白天开启之后，这些哈哈镜则变成了孩子们关注的焦点。

NORWEGIAN WILD REINDEER CENTRE PAVILION
挪威野生驯鹿中心
COPY EROSION 仿侵蚀结构

Location: Norway
Completion year: 2011
Designer: Snøhetta Oslo AS
Use: Watching Wildlife
Site Area: 75㎡
Client: Norwegian Wild Reindeer Foundation
Photographer: Ketil Jacobsen

项目地址：挪威
竣工时间：2011年
项目设计：斯诺赫塔建筑事务所
建筑用途：野外观察
项目面积：75 m²
开 发 商：挪威野生驯鹿基金会
摄 影 师：Ketil Jacobsen

Project Introduction

By appointment of the Norwegian Wild Reindeer Foundation, Snohetta has designed an observation and information pavilion at Hjerkinn in Dovre, Norway. The spectacular site is located on the outskirts of Dovrefjell National Park at around 1,250 meters above sea level, overlooking the Snohetta mountain massif. The main purpose of the 75㎡ building is to provide shelter for school groups and visitors as mountain guides lecture about the unique wildlife and history of the Dovre Mountain plateau.

项目介绍

TVERRFJELLHYTTA——挪威野生驯鹿中心，位于多夫勒山国家公园附近的荒原上，由挪威野生驯鹿基金会委任斯诺赫塔建筑事务所设计。建筑所在地海拔 1 250 m，可鸟瞰 Snohetta 山区。该中心将对公众开放，为野外观察和举办相关教育活动提供场所。

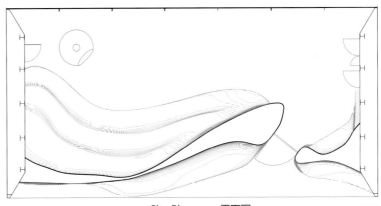

Site Plan —— 平面图　　PLAN

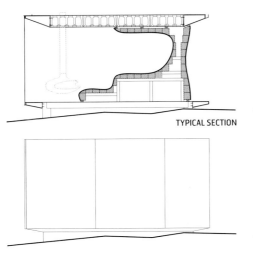

TYPICAL SECTION

WEST FACADE

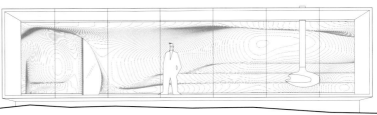

South Elevation —— 南立面　　SOUTH ELEVATION

North Elevation —— 北立面　　NORTH ELEVATION

Design Conception

Natural, cultural and mythical landscapes form the basis of the architectural idea. The building design is based on a contrast between a rigid outer shell and a soft organic-shaped inner core. A wooden core is placed within a rectangular frame of raw steel and glass. The core is shaped like rock or ice is eroded by natural forces like wind and running water. Its shape creates a protected and warm gathering place, while still preserving visitor's access to spectacular views.

设计构思

自然环境、区域文化和奇特的地貌,为建筑设计带来了灵感。棱角分明的立面和内部柔和的有机结构,产生了强烈的对比效果。这个结构类似受外力侵蚀后的石头或冰块,它为游客提供了安全、温暖的休憩之处的同时,丝毫不影响室外景色的观赏效果。

YUSUHARA WOODEN BRIDGE MUSEUM
梼原町木桥博物馆
LINKING TRADITIONAL AND CONTEMPORARY EXPRESSIONS 实现传统与现代的对话

Location: Kochi, Japan
Completion year: 2009
Designer: Kengo Kuma & Associates
Use: Exhibition Space
Site Area: 14,736.47 m²
Client: Tomio Yano, Town Mayor of Yusuhara
Photographer: Takumi Ota Photography

项目地址：日本高知县
竣工时间：2009 年
项目设计：隈研吾建筑事务所
建筑用途：展览空间
项目面积：14 736.47 m²
开 发 商：Tomio Yano, Town Mayor of Yusuhara
摄 影 师：太田拓实写真事务所

Project Introduction

This is a plan to link two public buildings with a bridge-typed facility, which had been long separated by the road between them. The museum technically bridges communications in this area. It functions not only as a passage between the two facilities but also as an accommodation and workshop, ideal location for artist-in-residence programs.

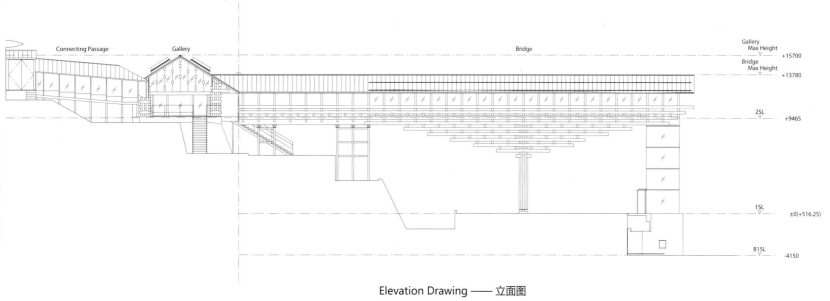

Elevation Drawing —— 立面图

North Elevation 1:300(A4)

项目介绍

这是一个连接被马路远远隔开的两处公共建筑物的桥型博物馆。它不仅从技术上搭建了沟通的桥梁，同时还是功能性的设施，是理想的工作间和艺术家的居住地。

Design Conception

In this project, we challenged a structural system which composes of small parts, referring to cantilever structure often employed in traditional architecture in Japan and China. It is a great example of sustainable design, as you can achieve a big cantilever even without large-sized materials. Blending traditional Japanese aesthetics with a contemporary language, the museum seeks to harmoniously coexist with its surrounding natural landscape.

设计构思

在这个项目中,设计师借用日本与中国传统建筑都常用的结构体"斗拱",实现了用小元件搭建结构系统的挑战。这是一个很成功的结构范例,它证明了不依赖大型的材料,也能建造巨大的悬臂。倒三角的造型很引人注目,它将日本的传统美学与当代建筑互相紧密结合起来,让建筑与周边自然景观和谐共处。

ONE OCEAN
2012 丽水世博会主题亭
WAVING STRUCTURE 流动的海洋形态

Location: Yeosu, South-Korea
Completion year: 2012
Designer: Soma
Use: Thematic Pavilion
Site Area: 8,200 m²
Client: Yeosu World Expo Preparatory Working committee
Photographer: Soma

项目地址：韩国丽水
竣工时间：2012 年
项目设计：Soma
建筑用途：展览
项目面积：8 200 m²
开 发 商：丽水世博会筹委会
摄 影 师：Soma

Project Introduction

The main design intent was to embody the Expo's theme The Living Ocean and Coast and transform it into a multi-layered architectural experience. Therefore the Expo's agenda, namely the responsible use of natural resources was not visually represented, but actually embedded into the building, snch as through the sustainable climate design or the biomimetic approach of the kinetic façade. The cutting-edge façade system was developed together with Knippers Helbig Advanced Engineering and supports the aim of the world exhibition to introduce forward-looking innovations to the public.

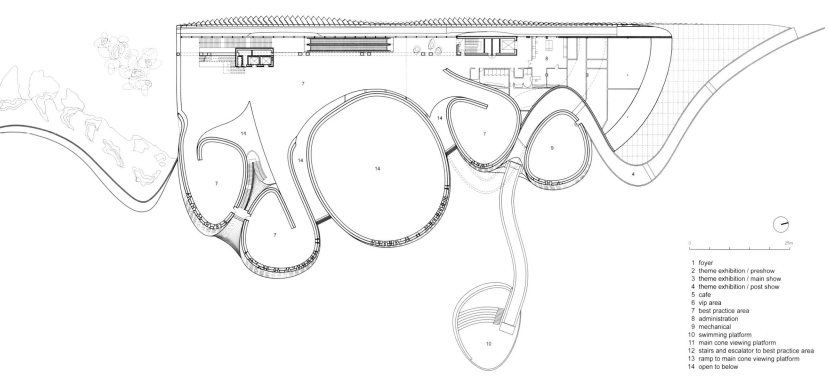

1 foyer
2 theme exhibition / preshow
3 theme exhibition / main show
4 theme exhibition / post show
5 cafe
6 vip area
7 best practice area
8 administration
9 mechanical
10 swimming platform
11 main cone viewing platform
12 stairs and escalator to best practice area
13 ramp to main cone viewing platform
14 open to below

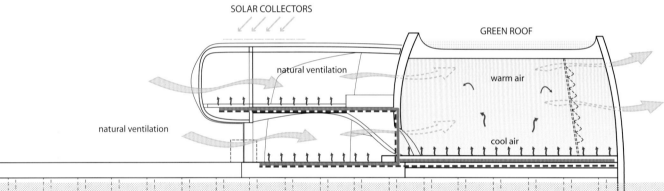

项目介绍

这个展亭将包含两种不同的展示风格，它为游客介绍丽水世博会的主题及生动的海洋和海岸环境。

Design Conception

The designers experience the Ocean mainly in two ways, as an endless surface and an immersed perspective as depth. This plain/profound duality of the Ocean motivates the building's spatial and organizational concept. Continuous surfaces twist from vertical to horizontal orientation and define all significant interior spaces. The vertical cones invite the visitor to immerse into the Thematic Exhibition. They evolve into horizontal levels that cover the foyer and become a flexible stage for the Best Practice Area.

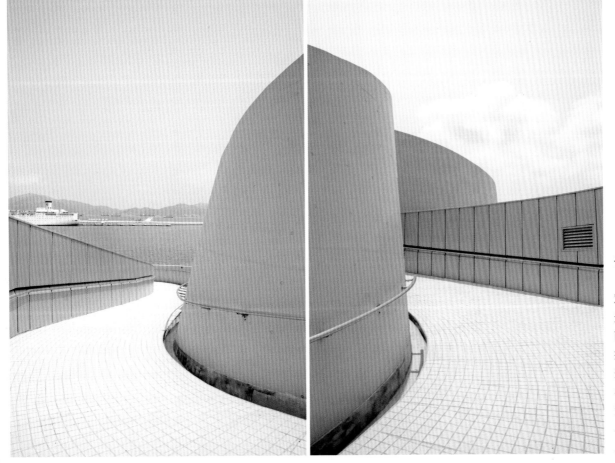

设计构思

对于海洋的体验可以分为两种主要的形式，一是无限的水平，另一个是融入其中的深度体验，这两种形式通过建筑的空间和组织形式展现出来。持续的表层因为垂直又水平的定位而产生扭曲变形，从而丰富了室内空间。垂直的主体部分将游客引入内部的主题展示。垂直的主体部分朝着海洋，象征弯曲的海岸线；柔和的边界形成了海水与陆地之间的持续对比；展亭的另一侧形成了人工屋顶景观，包含花园和道路。

ROCK IT SUDA
ROCK IT SUDA 度假屋
COLORFUL RAINBOW 摇滚彩虹

Location: Ganwon do, Korea
Completion year: 2009
Designer: Moon Hoon
Use: Weekend House
Site Area: 456.39 m²
Client: Kim Jae il
Photographer: Kim yong kwan

项目地址：韩国江原道
竣工时间：2009 年
项目设计：Moon Hoon
建筑用途：周末度假屋
项目面积：456.39 m²
开 发 商：Kim Jae il
摄 影 师：Kim yong kwan

Project Introduction

Rock It Suda is a name for the pension and its concept. The client is a bass guitarist, a member of the amateur rock group named Rock It Suda. On regular basis, they perform live, at the pension. The site has a dry river, and an open view in front and the other sides are surrounded by the mountain range.

项目介绍

在韩国江原道茂密的森林中，隐藏着一组色彩艳丽的异形建筑，这就是由 Moon Hoon 设计的 Rock It Suda 度假屋。Rock It Suda 这个名字源自于业主乐队名字，作为一名业余的贝司手，对音乐和艺术的热爱让他和设计师一拍即合，创造了一组这样疯狂的建筑。这个度假村，紧靠一条干涸的小河，屋前视野极好，背靠绵延的群山。

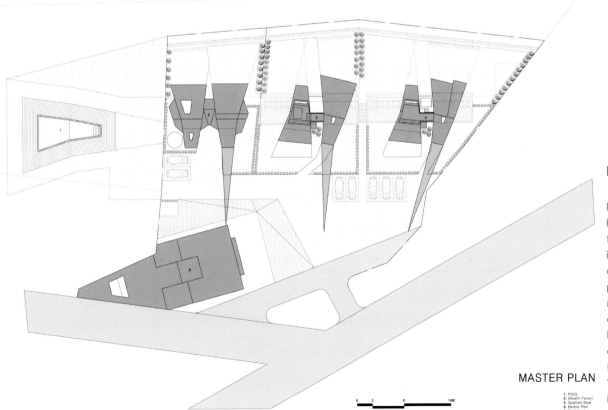

Panoramic View Plan —— 全景平面图

Design Conception

Play architecture is born, when giant soft hammock, shaped like tails are attached to hard secured architecture. The clever invention not only adds to the poetic, dynamic nature of the complex but also portrays the pension as a living organism. Moon gave each volume a clearly discernible identity – car, plane, bull, Barbie doll – accompanied by a vibrant colour that gives the unit a name: Ferrari Red, Stealth Black, Spanish Blue, Flamenco White, Barbie Pink and Oriental Gold. It is hoped that, the wind, and movement made by people playing in the hammock expresses the desire that architecture is not secure and unmoving.

MASTER PLAN

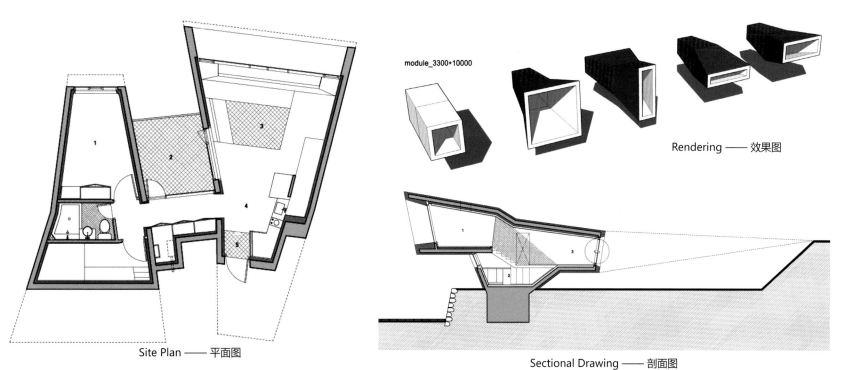

Site Plan —— 平面图

Rendering —— 效果图

Sectional Drawing —— 剖面图

设计构思

作为 Moon Hoon "玩建筑"的又一次大胆尝试，他将独特的个人风格与丰富多彩的精神世界注入冰冷的建筑之中，让建筑有了自己性格和色彩，让它们"活"了起来。他从其他艺术领域汲取灵感来丰富设计构想，为每个建筑都设计了独立的空间主题，最终达到了多种文化的融合。设计师从客人的不同喜好出发，选取了芭比娃娃、西班牙斗牛、隐形战斗机和法拉利跑车等主题。每个建筑外观都有相应主题的配色，黄色、粉红、蓝色、黑色等，色彩浓烈。设计师通过对长方体的变形，构成了收缩、扩张等空间形式，创造出一个个极富个性的象征式建筑。

ROOFTECTURE HH
ROOFTECTURE HH 私人住宅
SURROUNDED BY CHERRY 樱花深处有人家

Location: Hyogo, Japan
Completion year: 2010
Designer: Endo Shuhei Architect Institute
Use: House
Site Area: 1,175 m²

项目地址：日本兵库县
竣工时间：2010 年
项目设计：Endo Shuhei Architect Institute
建筑用途：住宅
项目面积：1 175 m²

Project Introduction

This project is a private residence which located in a lush which is surrounded by mountains in western Hyogo Prefecture of Japan. There are 4 large cherry trees planted in the center of this site which beautiful blooms every year. One moment of the nature scale created a powerful landscape beyond any architectural peculiarities space. The strong power also is the originality of this architecture.

项目介绍

该项目是一个私人住宅,坐落于日本兵库县西部的群山上,四周郁郁葱葱。四棵巨大的樱花树种在场地的中央,年年绽放,每每花瓣如粉红瀑布般悬挂而下,极富诗情画意。建筑上的独创性与独特的自然美景,一同赋予这个空间别样的美丽。

1 Entrance
2 Piano room
3 Dining room
4 Kitchen
5 Living room (Japanese style room)
6 Guest room (Japanese style room)
7 Bed room (Japanese style room)
8 Toilet
9 Wash room
10 Bath room
11 Strage

Site Plan / Plan S=1/100

Site Plan —— 平面图

North Elevation——北立面

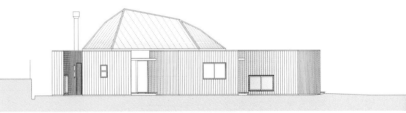

South Elevation——南立面

Design Conception

This residence was planned as a second house for the client who can enjoy a relaxed life and live with those lovely cherry blossoms. Client asked to views cherry blossoms from inside of the house, and the living space should be created as to feel the light and wind. Functions of housing are dispersed in four wooden boxes, staggered angles for each direction respectively to views the mountains and cherry around. Living room is setting in the center of this house where surrounding by those 4 boxes. Living room is framed as a big roof which covered by a translucent glass fiber-reinforced panels, hard to believe there is an indoor space because it is so brightness.

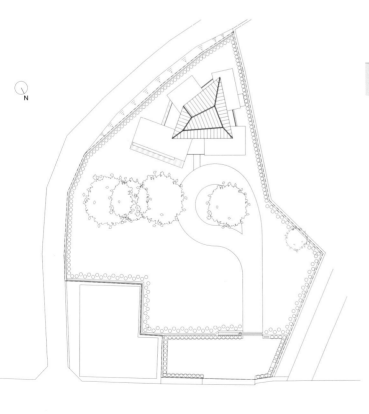

Panoramic View Plan —— 全景平面图

设计构思

该住宅共两层,为住户提供了舒适的空间和极佳的赏花视野。住户希望在室内也能欣赏到樱花缤纷落下的场景,还希望客厅可以有充足的采光和良好的通风。房屋的功能被分配在4个木制的"盒子"里。这4个"盒子"错列分布,角度不一,可以分别观赏到不同的景致。客厅被放置在了整个房子的中央,被4个盒状结构包裹。它被巨大的屋顶框住,覆盖着半透明的玻璃质强纤维面板。视觉上的明亮度,让人们很难想象,屋顶下还存在着一个室内空间。

THE WELLINGTON ZOO HUB
DINING COURTYARD
惠灵顿动物园活动中心 野餐庭院

Location: New Zealand
Completion year: 2011
Designer: Assembly Architects Limited
Use: Picnics
Client: Naylor Love Limited
Photographer: Jet Productions Ltd

项目地址：新西兰
竣工时间：2011年
项目设计：Assembly Architects Limited
建筑用途：野餐
开发商：Naylor Love Limited
摄影师：Jet Productions Ltd

Project Introduction

The Hub at Wellington Zoo, New Zealand features a diverse programme to suit the reptile inhabitants and the human visitors. Assembly Architects Limited repurposed the curious entry facade of a former elephant house to form a walled dining courtyard, used site excavations to form a rammed earth reptile habitat and designed a unique event pavilion for daytime picnics and evening parties.

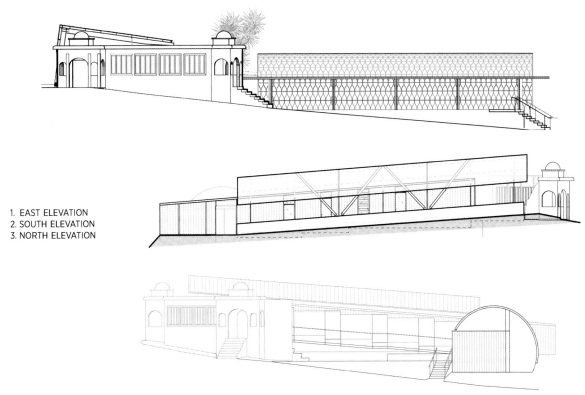

1. EAST ELEVATION
2. SOUTH ELEVATION
3. NORTH ELEVATION

Elevation Drawing —— 立面图

该项目坐落在市区临山处的绿地之中。它的前身是大象园，改建后变成一个有围墙的游客用餐庭院，同时也是爬行动物的栖息地。具体来说，设计团队将考古区打造成一个爬行动物的居住区，同时增加了另一个独特的亭子结构，用于白天的野餐和晚上的聚会。

项目介绍

该项目坐落在市区临山处的绿地之中。它的前身是大象园，改建后变成一个有围墙的游客用餐庭院，同时也是爬行动物的栖息地。具体来说，设计团队将考古区打造成一个爬行动物的居住区，同时增加了另一个独特的亭子结构，用于白天的野餐和晚上的聚会。

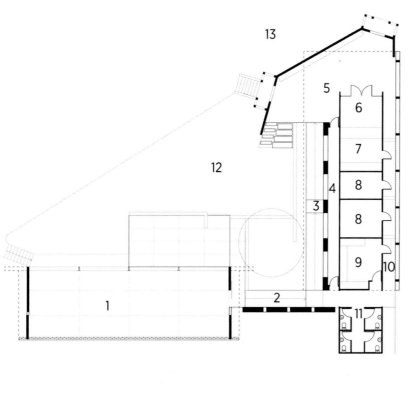

Site Plan —— 平面图

Design Conception

Almost 7m and length of 21m, the pavilion is with a clear span. The structure comprises a concrete slab, two precast concrete end walls, and a steel frame to support a vaulted tent structure. Extensive and collaborative design and prototyping was carried out for the vault framework. The final solution was thoroughly tested with up to 1:1 prototyping prior to final confirmation. The 4mm aluminium shape, fabricated and assembled by Fraser Engineering was cut on a turret punch. The shape was then pressed into the radius of the final vault on a form press, and powder-coated in gloss black paint. On site, following a precise set-out and erection of the steel frames, Fraser Engineering assembled 716 of the curved elements, weighing 1.5Kg per piece, working horizontally to position in place with around 1400-1500 turned aluminum rods through the pre punched holes. Finally, 1400-1500 tube gussets were riveted to the curved elements to balance the torsional forces in the structure and complete the vault framework. The pavilion assembly took approximately 3 weeks. The pavilion is clad in a lightweight plastic membrane. The opaque film lends a bright translucency to the pavilion interior, with interesting effects in both rain and sun. When the stunning late-afternoon shadow play from the western trees wanes, artificial lighting takes over and the effect is reversed, illuminating the form of the structure to the exterior at night.

设计构思

亭子长 21 m，跨度为 7 m，圆顶帐篷式的结构包括一块混凝土板，两块预制混凝土端墙及一个钢铁骨架。弧形外壳经过一系列的合作设计，做了 1:1 的足尺模型之后，由结构工程师敲定。4 mm 厚的铝条外覆有黑色烤漆，形成弯曲优美的造型，以单元进行组合，每个单元大约重 1.5 kg，共 716 个单元。该项目用到了 1 400~1 500 个管状金属充当锚件，从而平衡结构扭转力。整个框架完成大约耗费了 3 个星期。亭子外面笼罩着轻质塑料薄膜。半透明表皮在下午的时候树影婆娑，拥有让人流连忘返的光影；到了晚上，灯光亮起，这里内外灯火通明。

AKRON ART MUSEUM
阿克伦城艺术博物馆
FLOW OUT OF THE BUILDING 由内而外的艺术气息

Location: Ohio, USA
Completion year: 2007
Designer: Coop Himmelb(l)au
Use: Exhibition
Site Area: 8,370 m²
Client: Akron Art Museum
Photographer: Gerald Zugmann

项目地址：美国俄亥俄州
竣工时间：2007 年
项目设计：蓝天组建筑事务所
建筑用途：展览
项目面积：8 370 m²
开 发 商：阿克伦城艺术博物馆
摄 影 师：Gerald Zugmann

Project Introduction

The building is diveded into 3 parts: the Crystal, the Gallery Box, and the Roof Cloud.

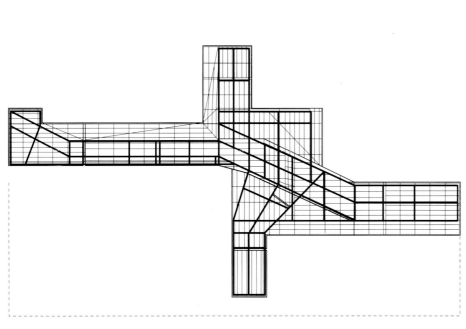

Site Plan —— 平面图

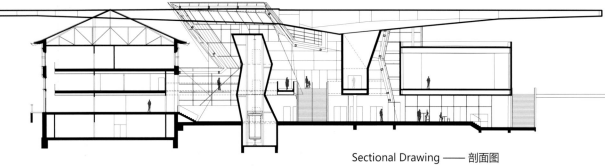

Sectional Drawing —— 剖面图

项目介绍

该美术馆由三个部分组成：水晶宫（Crystal）、艺廊区块（Gallery Box）和云顶（Roof Cloud）。

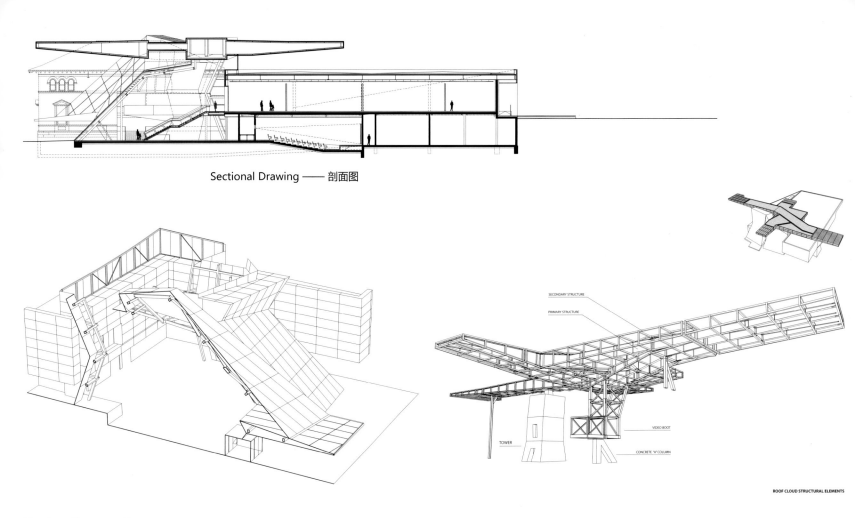

Sectional Drawing —— 剖面图

Design Conception

The Crystal serves as the main entry and operates as an orientation and connection space serving both the new and old buildings. The traditional idea of a banquet hall as an enclosed isolated event space dissolves away into a visible, public experience. The Roof Cloud, which hovers above the building, creates a blurred envelope for the museum because of its sheer mass and materiality. It encloses interior space, provides shade for exterior spaces, and operates as a horizontal landmark in the city.

设计构思

水晶宫作为美术馆的主入口,不仅是一个指示方位的场所,还与新旧建筑物的公共功能区(如礼堂、教室、图书馆、咖啡馆、书店)相连接。宴会厅作为一个独立场所,将变为一种可视化的公共体验。云顶盘旋在该建筑物的上方,给美术馆增添了一层朦胧的外壳。它具有内部空间,在该建筑物的外部形成阴影区域,同时也将成为该城市的水平地标。从 Akron 市区大道或会议中心,甚至更远处,都可以看到标示美术馆所在的云顶。美术馆的一些活动,如现场音乐会和露天宴会,都可以在云顶下方的城市雕塑公园或地面层的活动场所举行。

BMW WELT
宝马世界
FASCINATING INSIGHTS 全方位的感受

Location: Munich, Germany
Completion year: 2007
Designer: Coop Himmelb(l)au
Use: Multi-purpose
Site Area: 25,000 m²
Client: BMW AG
Photographer: Ari Marcopoulos

项目地址：德国慕尼黑
竣工时间：2007年
项目设计：蓝天组建筑事务所
建筑用途：多功能
项目面积：25 000 m²
开 发 商：宝马汽车公司
摄 影 师：Ari Marcopoulos

Project Introduction

Located in Munich within close proximity to BMW headquarters, their main manufacturing plant and the BMW museum, BMW Welt is home to automobile exhibitions, an automobile delivery centre and an event hub all under one roof.Customers and visitors are provided with a closer than ever insight into all aspects of the BMW brand.

项目介绍

该项目设在德国慕尼黑的宝马世界,与 BMW 总部、BMW 总工厂及 BMW 博物馆毗邻而居,它不仅是伟大的建筑作品,也是 BMW 新车交付中心和品牌体验中心。对于来自全球的客户和参观者而言,没有比这里更能展现 BMW 魅力的地方了。

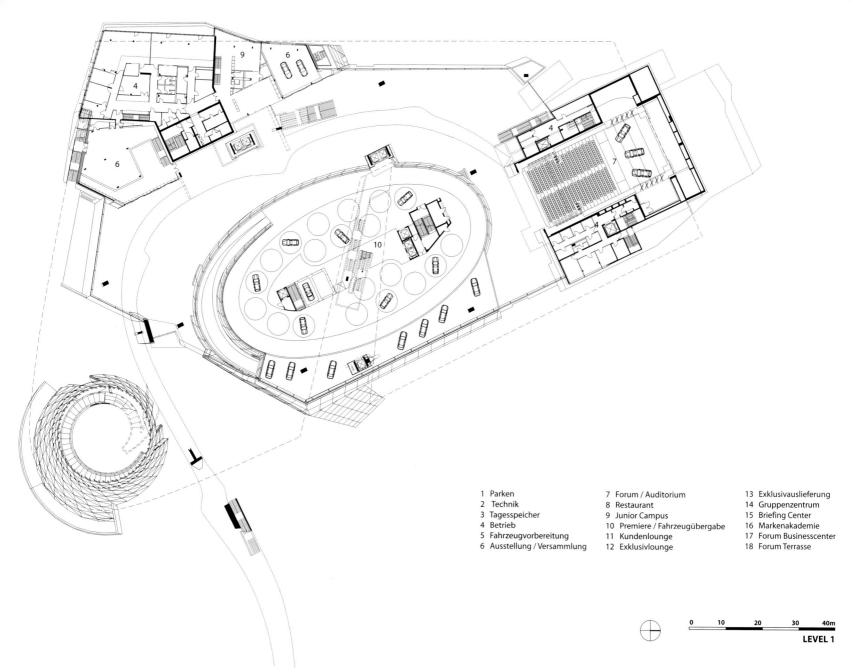

1 Parken	7 Forum / Auditorium	13 Exklusivauslieferung
2 Technik	8 Restaurant	14 Gruppenzentrum
3 Tagesspeicher	9 Junior Campus	15 Briefing Center
4 Betrieb	10 Premiere / Fahrzeugübergabe	16 Markenakademie
5 Fahrzeugvorbereitung	11 Kundenlounge	17 Forum Businesscenter
6 Ausstellung / Versammlung	12 Exklusivlounge	18 Forum Terrasse

LEVEL 1

Panoramic View Plan —— 全景平面图

Design Conception

The roof landscape of 16,000 square meters, creating the impression that it is floating, supported by only 12 columns, not only forms the space-enclosing upper limit of the building, but also forms a functional, structural, and above all formally independent structure, in conjunction with the Double Cone. The BMW Welt is probably the biggest 'blob-shaped' building. it is the result of an integrated design process by using building information modeling (BIM).

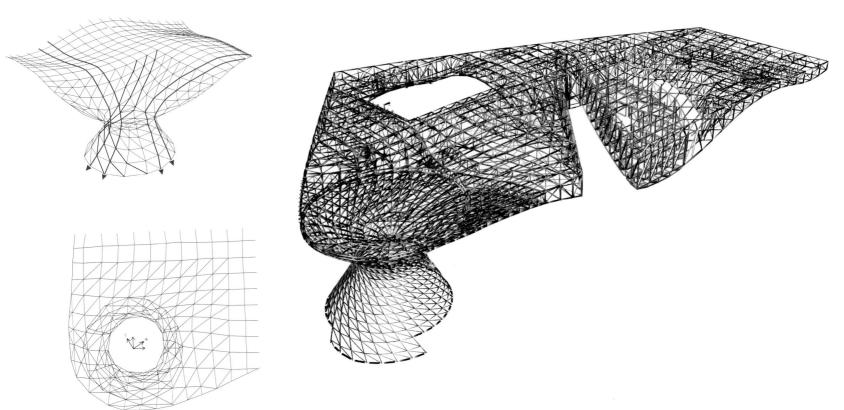

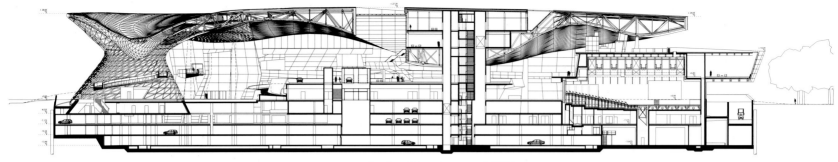

Sectional Drawing —— 剖面图

设计构思

一座 16 000 m² 的巨大屋顶宛若一片云彩，划入浅蓝色的天际。尽管有 10 个雅典卫城那么大，但屋顶只由 12 根铰链柱和一个令人惊叹的双圆锥造型钢结构进行支撑。整座建筑气势宏伟，却完全依靠浑然一体的各个部件傲然屹立。面对如此复杂的造型，设计师借助了建筑信息模型（BIM：Building Information Modeling）的理念，应用 Autodesk Revit 三维软件设计而成。

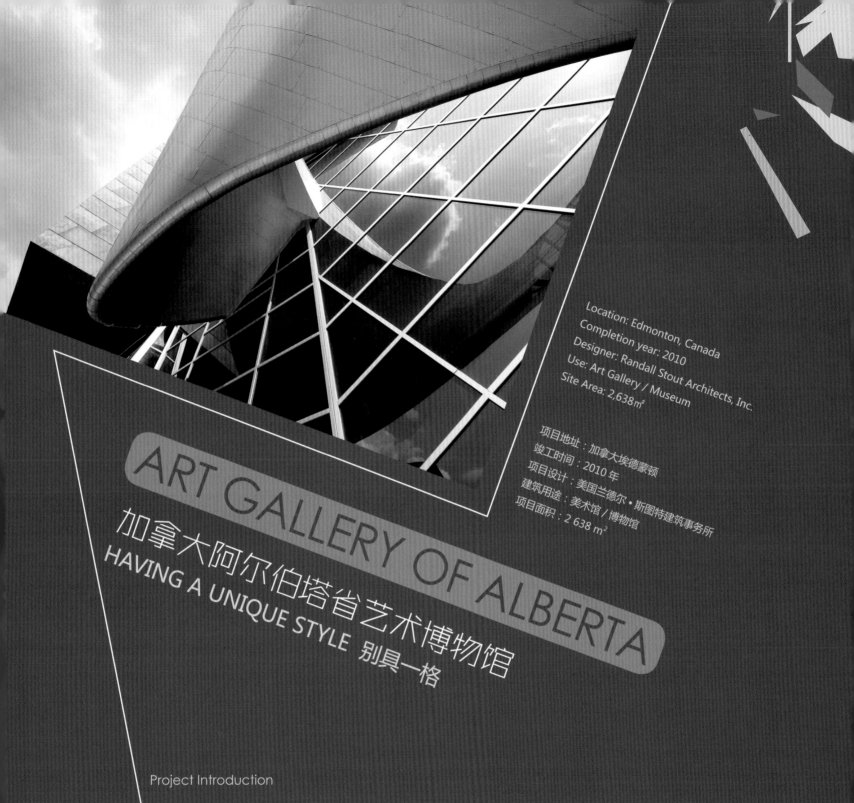

ART GALLERY OF ALBERTA
加拿大阿尔伯塔省艺术博物馆
HAVING A UNIQUE STYLE 别具一格

Location: Edmonton, Canada
Completion year: 2010
Designer: Randall Stout Architects, Inc.
Use: Art Gallery / Museum
Site Area: 2,638m²

项目地址：加拿大埃德蒙顿
竣工时间：2010 年
项目设计：美国兰德尔·斯图特建筑事务所
建筑用途：美术馆/博物馆
项目面积：2 638 m²

Project Introduction

Unlike most buildings in the harshly cold climate of northern Alberta, the new Art Gallery of Alberta expresses an engaging and stimulating social presence. Celebrating its prominent Edmonton location on Sir Winston Churchill Square, the main civic and public arts square in the city; the project represents the institution's commitment to enhancing the public's experience of the visual arts. The new Gallery includes an addition / renovation component that upgrades existing below-standard facilities and adds new celebratory public event areas that bring a new architectural vitality to Edmonton's urban core.

项目介绍

该项目与阿尔伯塔省北部的其他建筑有所不同,充满现代建筑艺术的迷人魅力,令人耳目一新。它位于埃德蒙顿最主要的公共艺术广场——温斯顿·丘吉尔广场,不仅丰富了人们的视觉体验,同时也体现出该机构的建设初衷。该项目在原有建筑的基础上进行翻新改造,升级原本低标准设施的同时也增加了新的公共活动空间,从而为市民营造了一个真正富有活力的活动空间。

Design Conception

Crafted of patinaed zinc, high performance glazing, and stainless steel, the building has a timeless appearance and extraordinary durability in the northern climate. Transparent glazing planes and reflective metal surfaces animate of the building, exposing the activities within and engaging people and art at multiple levels on both the interior and exterior. Not only does the building change throughout the day, it also changes from season to season.

设计构思

氧化锌、高性能玻璃、不锈钢，不仅赋予建筑时尚的外观，也是针对北方气候提升建筑耐久性最好的材料。透明的玻璃平面和具有反射能力的金属外观熠熠生辉，使得建筑内外的景象与活动一目了然。该建筑不仅会随着一日光景的变化而变化，还会随季节的变化而变化。

TROLLWALL RESTAURANT
TROLLWALL 饭店
WONDERFUL ARTICAL EXCELLING NATURE 巧夺天工

Location: Møre og Romsdal, Norway
Completion year: 2011
Designer: Reiulf Ramstad Arkitekter (RRA)
Use: Restaurant and Service Building
Site Area: 700 m²
Client: Private
Photographer: Reiulf Ramstad Architects

项目地址：挪威默勒 - 鲁姆斯达尔
竣工时间：2011 年
项目设计：Reiulf Ramstad Arkitekter (RRA)
建筑用途：饭店和服务大楼
项目面积：700 m²
开 发 商：Private
摄 影 师：Reiulf Ramstad Architects

Project Introduction

It's a new cursor at the foot of the Troll Wall. The architecture of the new visitors' center next to E139 is an outcome of the sites close connection to the impressive mountain wall, which is Europe's tallest vertical, overhanging rock face in The Romsdal Valley.

项目介绍

因为位于著名的 Troll Wall 山脚下，该建筑成为了市区的标志性建筑和新的观光胜地。Troll Wall 是欧洲最高的山。该建筑突出的部分延伸到了 Romsdal 山谷。

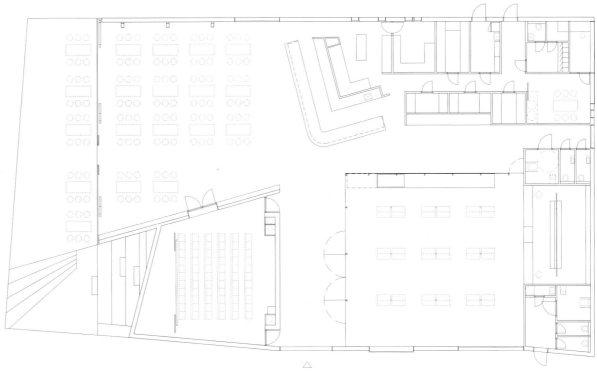

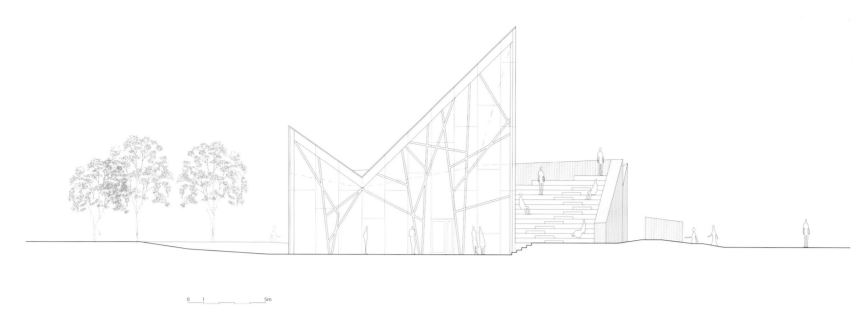

Design Conception

It is building a character and identity which in itself will be an attraction in the region. The building has a simple and flexible plan, with a characteristic roof that has its character from the majestic surrounding landscape. This new building has a dramatically sloped roof that mimics its surroundings with eye-catching glass façade to reflect the impressive natural landscape. The M-shaped glass roof really looks fantastic both from the inside and the outside.

设计构思

该建筑在该地区不仅辨识度高,且极富吸引力。构思简单,设计巧妙,其极具特色的屋顶与周遭宏伟壮丽的山色融为一体。极度倾斜的屋顶模仿了周围的环境。玻璃立面的采用使得四周美丽的自然风光被映射其中,显得耀眼而夺目。无论从里面还是外面看,M 形状的玻璃屋顶的设计都称得上巧夺天工。

BUSAN CINEMA CENTER
韩国釜山电影中心
MOVIE PALACE 映画殿堂

Location: Busan, South Korea
Completion year: 2012
Designer: Coop Himmelb(l)au
Use: Multi-purpose
Site Area: 32,100 ㎡
Client: Municipality of Busan
Photographer: Markus Pillhofer

项目地址：韩国釜山
竣工时间：2012年
项目设计：蓝天组建筑事务所
建筑用途：多功能
项目面积：32 100 ㎡
开 发 商：釜山市政府
摄 影 师：Markus Pillhofer

Project Introduction

Busan Cinema Center, the new home of the Busan Film Festival (BIFF), combines open space, cultural program, entertainment, technology and architecture in a novel way.

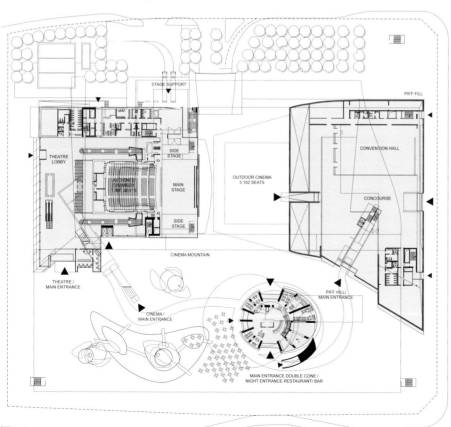

Panoramic View Plan —— 全景平面图

项目介绍

该项目是釜山国际电影节的新场地，它以一种创新的方式，将开放式空间、文化、娱乐、科技和建筑等元素融为一体。

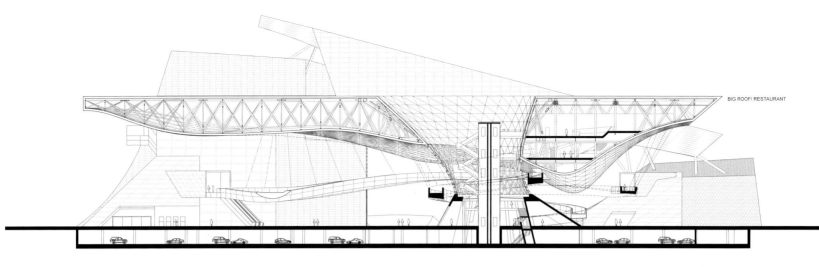

Sectional Drawing —— 剖面图

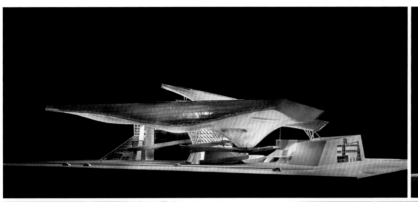
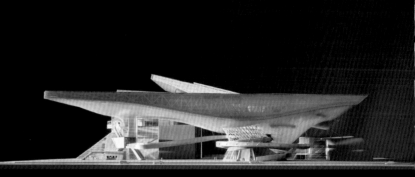

Design Conception

The dynamic LED lighting surface covering the undulating ceilings of the outdoor roof canopies gives the Busan Cinema Center its symbolic and representative iconographic feature. Artistic lighting programs tailored to events of the BIFF or the Municipality of Busan can be created by visual artists and displayed across the ceiling in full motion graphics, creating a lively urban situation at night, but if it's necessary it also can be visible during the day.

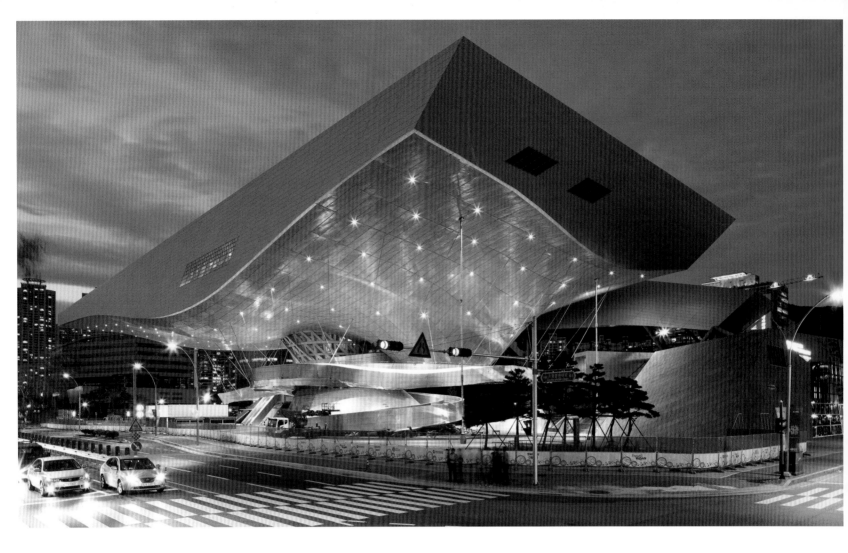

设计构思

波浪状的户外屋顶天蓬上,覆盖着动态 LED 灯幕,为建筑打造出一个极具标示性的外观。艺术性照明程序可以根据釜山国际电影节及釜山市政府的需要,预制动态画面。它不但可以在夜晚为城市注入活力,白天也同样可以显示。

WESTEND GATE
万豪酒店
TREE STRUCTURE 树状结构

Location: Frankfurt, Germany
Completion year: 2010
Designer: Just Burgeff architekten
Use: Hotel
Site Area: 1,000 m²
Client: Aberdeen GmbH + Frankfurt am Main
Photographer: Eibe Sönnecken

项目地址：德国法兰克福
竣工时间：2010 年
项目设计：Just Burgeff architekten
建筑用途：酒店
项目面积：1 000 m²
开 发 商：Aberdeen GmbH + Frankfurt am Main
摄 影 师：Eibe Sönnecken

Project Introduction

The organic tree-like structure of the sculptural roof is easily recognizable from a great distance. 1 000 square meters of roof surface extend at one point up to a height of 14 meters, zoning in the square, with translucent air cushions filling out the construction and providing office workers and hotel guests refuge from the rain.

项目介绍

即使站在远处,屋顶的有机树造型也极其抢眼。屋顶总面积为 1 000 m²,高 14 m。半透明的气垫遮雨蓬丰富了整个结构,为办公人员和酒店客人提供了绝佳的避雨场所。

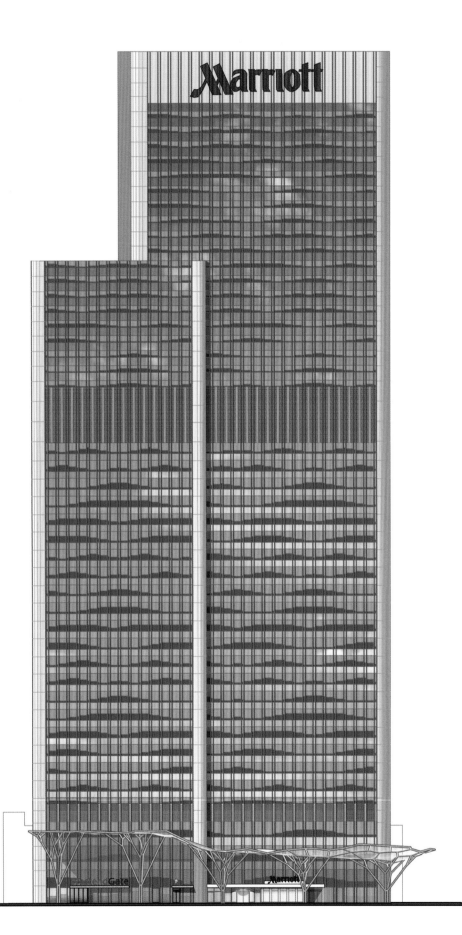

Elevation Drawing —— 立面图

Design Conception

The double curved, hive-like roof surfaces were fitted with transparent pneumatic ETFE plastic cushions in different shapes. Air supply hoses and cables were located inside the gutters and covered with grating. The permanently inflated cushions furnish protection against sun and rain in spite of their optical and physical airiness, offer an unencumbered view of the future new WestendGate façade.

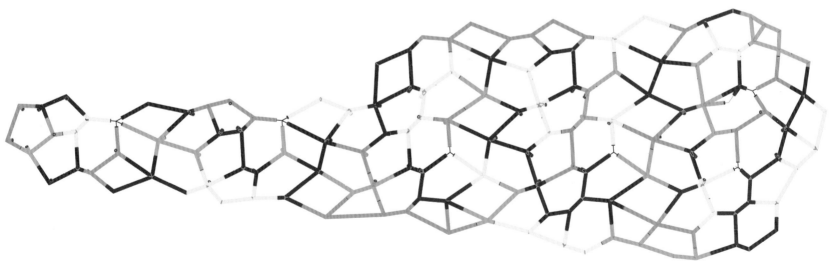

设计构思

双曲度、蜂窝造型的屋面配以形状不一、透明外观的气动 ETFE 塑料衬垫，使得该项目别具特色。供气软管和电缆被置于项目内部的水槽中，外面以箅子板覆盖。内部视野通透、空气流通自由。此外，长期处于膨胀状态的衬垫能有效阻挡阳光照射、防止雨水侵蚀。这使得未来西区大门全新的景观指日可待。

RIZHAO LANDSCAPING PROJECT
日照山海天阳光海岸配套公建
NATURALLY MOULDED 自然成趣

PLocation: Shandong, China
Completion year: 2012
Designer: HHD FUN
Use: Ancillary Facility

项目地址：中国山东
竣工时间：2012 年
项目设计：HHD FUN
建筑用途：配套公建

Project Introduction

HHD FUN, the architects selected for the transformation of a 2.5 km stretch of RiZhao sea coast have designed a post modernistic sculptured collection of buildings.

项目介绍

该项目是一项 2.5 km 的日照海岸改造工程，它由一系列具有城市装饰作用的建筑组成，极具现代感。

Design Conception

The architecture boasts the naturally moulded structures, with a minimalist and modest approach to the use of building matter. The structures are designed to integrate into the existing landscape, embedding and extending the natural contours present on site. Each individual unit will be constructed with natural roof lighting as well as a several units consisting of light permeable facades and facades equipped for exterior lighting, which will be used to lighten the paths.

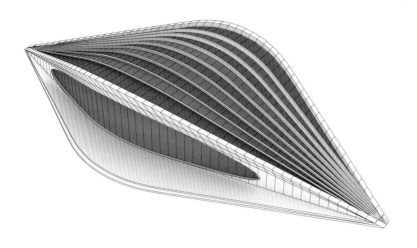

设计构思

该项目以自然成型的结构自诩,简约、内敛。建筑和景观水乳交融,丰富了区域的自然景观。由于大部分建筑都需要保证使用者的私密性,所以建筑形态相对封闭。房间从屋顶自然采光或通过立面透光。为了丰富海岸的夜景,有些立面还设置了 LED 灯为户外小径照明。

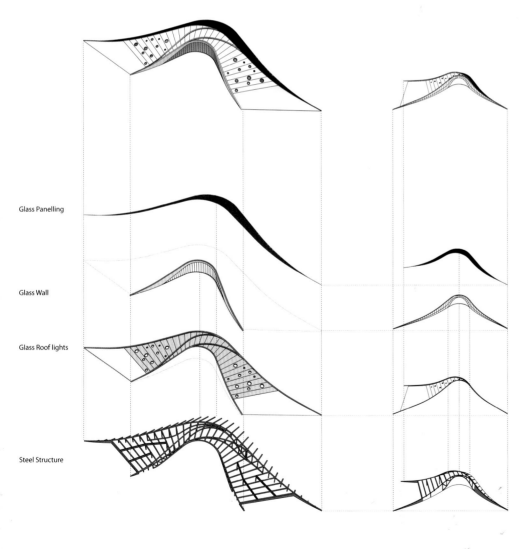

Parametricly designed structure

Variation of structure and design used for shower and toilet facilities

Rizhao Club House —— 日照俱乐部结构分析图
Structure Analysis

图书在版编目（CIP）数据

城市建筑装饰 / 香港理工国际出版社 主编 . – 武汉 : 华中科技大学出版社 , 2012.8
ISBN 978-7-5609-8287-8

Ⅰ . ①城… Ⅱ . ①香… Ⅲ . ①建筑装饰 – 建筑设计Ⅳ . ① TU238

中国版本图书馆 CIP 数据核字（2012）第 182009 号

城市建筑装饰	香港理工国际出版社 主编

出版发行：华中科技大学出版社（中国·武汉）
地　　址：武汉市武昌珞喻路1037号（邮编：430074）
出 版 人：阮海洪

责任编辑：段自强	责任监印：秦英
责任校对：段园园	装帧设计：百彤文化

印　　刷：利丰雅高印刷（深圳）有限公司
开　　本：992 mm × 1240 mm　1/16
印　　张：19.75
字　　数：158千字
版　　次：2013年3月第1版 第1次印刷
定　　价：288.00元（USD 57.99）

投稿热线：（020）36218949　　1275336759@qq.com
本书若有印装质量问题，请向出版社营销中心调换
全国免费服务热线：400-6679-118 竭诚为您服务
版权所有　侵权必究